SpringerBriefs in Mathematical Physics

Volume 38

SpringerBriefs are characterized in general by their size (50–125 pages) and fast production time (2–3 months compared to 6 months for a monograph).
Briefs are available in print but are intended as a primarily electronic publication to be included in Springer's e-book package.

Typical works might include:

- An extended survey of a field
- A link between new research papers published in journal articles
- A presentation of core concepts that doctoral students must understand in order to make independent contributions
- Lecture notes making a specialist topic accessible for non-specialist readers.

SpringerBriefs in Mathematical Physics showcase, in a compact format, topics of current relevance in the field of mathematical physics. Published titles will encompass all areas of theoretical and mathematical physics. This series is intended for mathematicians, physicists, and other scientists, as well as doctoral students in related areas.

Editorial Board

- Nathanaël Berestycki (University of Cambridge, UK)
- Mihalis Dafermos (University of Cambridge, UK / Princeton University, US)
- Atsuo Kuniba (University of Tokyo, Japan)
- Matilde Marcolli (CALTECH, US)
- Bruno Nachtergaele (UC Davis, US)
- Hirosi Ooguri (California Institute of Technology, US / Kavli IPMU, Japan)

Springer Briefs in a nutshell

SpringerBriefs specifications vary depending on the title. In general, each Brief will have:

- 50–125 published pages, including all tables, figures, and references
- Softcover binding
- Copyright to remain in author's name
- Versions in print, eBook, and MyCopy

More information about this series at http://www.springer.com/series/11953

Hiroshi Isozaki

Inverse Spectral and Scattering Theory

An Introduction

Hiroshi Isozaki
Department of Mathematics
Tsukuba University
Tsukuba
Ibaraki, Japan

ISSN 2197-1757 ISSN 2197-1765 (electronic)
SpringerBriefs in Mathematical Physics
ISBN 978-981-15-8198-4 ISBN 978-981-15-8199-1 (eBook)
https://doi.org/10.1007/978-981-15-8199-1

This Springer imprint is published by the registered company Springer Nature Singapore Pte Ltd.
The registered company address is: 152 Beach Road, #21-01/04 Gateway East, Singapore 189721, Singapore

Preface

1. Inverse Spectral Problems Waves propagate carrying characteristic features of surrounding media, more generally, ambient spaces. By observing the wave motion, one may seek geometric or physical properties of space even for the remote part, since waves are composed of spectral quantities that contain global information of the space. The aim of this book is a brief introduction to this basic problem:

The recovery of the operator and the space from the knowledge of solutions to wave equations observed on some local part of the space.

Let us pick up two typical examples. Consider a wave equation in a bounded domain Ω in $\mathbf{R}^n$:

$$\frac{\partial^2}{\partial t^2} u = -Hu = \left(\sum_{i,j=1}^{n} a_{ij}(x) \frac{\partial^2}{\partial x_i \partial x_j} + \sum_{i=1}^{n} b_i(x) \frac{\partial}{\partial x_i} + c(x) \right) u$$

with given boundary data: $u(t, x) = f(t, x)$ on $\partial\Omega$ and initial data $u(0, x) = u_0(x)$, $\frac{\partial}{\partial t} u(0, x) = u_1(x)$. A natural observation is the motion at the boundary, namely $\frac{\partial}{\partial \nu_A} u(t, x)\big|_{\partial\Omega} = \sum_{i,j=1}^{n} a_{ij} \nu_i \partial_j u\big|_{\partial\Omega}$, where $\nu = (\nu_1, \cdots, \nu_n)$ is the outer unit normal to $\partial\Omega$. By passing to the time-periodic wave, $u(t, x) = e^{i\sqrt{\lambda}t} u(x)$, the issue is reduced to a boundary value problem:

$$Hu = \lambda u \quad \text{in} \quad \Omega, \quad u(x) = f(x) \quad \text{on} \quad \partial\Omega. \tag{0.1}$$

One can then associate the *Dirichlet-to-Neumann map (D-N map)* defined by

$$\Lambda(\lambda) f = \frac{\partial u}{\partial \nu_A}\bigg|_{\partial\Omega}.$$

The problem we address is now formulated as:

Can one determine the coefficients of H from the knowledge of the D-N map?

Our another motivation is a motion of quantum mechanical particles scattered by a potential. Let $H = -\Delta + V(x)$ be a Schrödinger operator on $\mathbf{R}^n$ where the potential $V(x)$ is real-valued and decays sufficiently rapidly at infinity. Observing the wave propagation by the Schrödinger equation $i \frac{\partial}{\partial t} u = Hu$, we can define Heisenberg's *S-matrix* (or *scattering matrix*), which assigns the profile of the solution at infinity in the remote future to that in the remote past. The inverse problem of scattering is then formulated as follows:

Can one construct the potential $V(x)$ from the S-matrix ?

For the one-dimensional case, these problems were solved in a complete form in the 1950s. Remarkable progress for the multi-dimensional case appeared in the middle of 1960s and is still enlarging its frontiers. Since we have considerable knowledge of inverse spectral and scattering problems at present, to have an overview by recalling basic achievements that serve to form a framework of our understanding is worthwhile.

2. One-Dimensional Problems We consider an operator $-(d/dx)^2 + V(x)$ with a real-valued potential $V(x)$ on $(0, 1)$ under the Dirichlet boundary condition. It has the eigenvalues $\lambda_1(V) < \lambda_2(V) < \cdots < \lambda_n(V) < \cdots \to \infty$. We raise a simple problem.

If $\lambda_n(V_1) = \lambda_n(V_2)$ for all $n \geq 1$, then $V_1(x) = V_2(x)$?

The answer is easily seen to be negative if we take $V_2(x) = V_1(1 - x)$. The knowledge of the eigenvalues is not sufficient to determine the potential. Then, what kind of data shall we use in addition to the eigenvalues? It is the derivative at the boundary of eigenfunctions. Namely, letting $\varphi_n(x\,;V)$ be a normalized eigenvector associated with $\lambda_n(V)$, we can adopt

$$\{(\varphi_n'(0\,;V), \varphi_n'(1\,;V))\,;\, n = 1, 2, \cdots\}, \quad ' = d/dx$$

as the spectral data. As a matter of fact, it is enough to employ the Neumann data at one end point. Let $y(x, \lambda\,;V)$ be the solution of the Cauchy problem:

$$\begin{cases} \left(-\dfrac{d^2}{dx^2} + V(x)\right) y(x, \lambda\,;V) = \lambda y(x, \lambda\,;V), \quad \text{on} \quad (0, 1), \\ y(0, \lambda\,;V) = 0, \quad y'(0, \lambda\,;V) = 1. \end{cases}$$

The eigenvalues $\lambda_n(V)$ are the zeros of $y(1, \lambda\,;V)$. Let $\kappa_n(V) = y'(1, \lambda_n(V); V)$. The fundamental result of the one-dimensional inverse spectral theory is:

The potential $V(x)$ is determined by $\lambda_n(V)$ and $\kappa_n(V)$ for all $n \geq 1$.

The *characterization problem* is also solved. Namely, one knows the necessary and sufficient condition for sequences $\{\lambda_n\}_{n=1}^{\infty}$, $\{\kappa_n\}_{n=1}^{\infty}$ to be the spectral data of some potential $V(x)$. It can further be rephrased in the following form. For a given potential $q(x)$, let $\mathcal{M}(q)$ be the set of V's having the same Dirichlet eigenvalues with q. Then,

$\mathcal{M}(q)$ is an infinite dimensional real analytic manifold[1] and $(\kappa_1(V), \kappa_2(V), \cdots)$ is a global coordinate system on $\mathcal{M}(q)$.

Therefore, in the one-dimensional inverse boundary value problem, one can deform the potential $V(x)$ arbitrarily keeping the eigenvalues fixed.

As for the inverse scattering, consider the Schrödinger equation $(-d^2/dx^2 + V(x))u = k^2 u$ on a half-line $(0, \infty)$ with Dirichlet condition $u(0) = 0$. In this case, the S-matrix is an operator of multiplication by a function $S(k)$ with modulus 1. The solution of inverse scattering problem on the half-line is stated in the following way:

The potential $V(x)$ is uniquely reconstructed from the S-matrix $S(k)$ for all $k \geq 0$, negative eigenvalues $-k_1^2 > \cdots > -k_m^2$ and the associated normalizing constants $\kappa_1, \cdots, \kappa_m$. Moreover, there is a necessary and sufficient condition for a function of k on $(0, \infty)$ to be the S-matrix of a Schrödinger operator $-d^2/dx^2 + V(x)$.

We thus see that the one-dimensional inverse problem is solved including the characterization problem.

3. Multi-Dimensional Problems

In multi-dimensions, one can start with a similar fact. Consider the Dirichlet eigenvalue problem for $-\Delta + V(x)$ on a bounded domain $\Omega \subset \mathbf{R}^n$. Let $\lambda_1(V) < \lambda_2(V) \leq \cdots \to \infty$ be the eigenvalues[2] and $\{\varphi_j(x; V)\}_{j=1}^{\infty}$ the associated orthonormal system of eigenvectors. Then, similarly to the case of one-dimension, we have:

Two potentials $V_1(x)$ and $V_2(x)$ coincide if $\lambda_j(V_1) = \lambda_j(V_2)$ and $\frac{\partial}{\partial \nu}\varphi_j(x; V_1) = \frac{\partial}{\partial \nu}\varphi_j(x; V_2)$ on $\partial\Omega$ for all $j \geq 1$.

However, the isospectral deformation of the potential similar to the one-dimensional case is no longer possible in multi-dimensions. This is a kind of spectral rigidity and is a feature of multi-dimensional inverse spectral problems.

As in the case of one-dimension, it is more convenient to use the D-N map as the spectral data. On a bounded domain $\Omega \subset \mathbf{R}^n$ with boundary $\partial\Omega$, consider the Dirichlet problem:

$$\begin{cases} (-\Delta + V(x))u = 0, & \text{in} \quad \Omega, \\ u = f, & \text{on} \quad \partial\Omega. \end{cases} \tag{0.2}$$

We then have:

For $n \geq 2$, the D-N map $\Lambda : f \to \left.\dfrac{\partial u}{\partial \nu}\right|_{\partial\Omega}$ determines the potential $V(x)$.

Compared with (0.1), this corresponds to the case of one fixed energy λ. Therefore, in multi-dimensions, the information from one fixed energy is enough to determine the potential. This is also true for the inverse scattering problem.

---

[1] It will be explained in Sect. 1.5.
[2] The first eigenvalue is simple.

In $\mathbf{R}^n$ with $n \geq 2$, compactly supported potentials are reconstructed from the S-matrix of arbitrarily fixed positive energy.

Another remarkable feature is:

In $\mathbf{R}^n$ with $n \geq 3$, the S-matrix is extended on some complex manifold and there exists a representation formula of the potential in terms of the extended S-matrix.

4. Riemannian Manifolds There is also a progress for the Laplace operator $H = -\sum_{i,j} \frac{1}{\sqrt{g}} \partial_i \left(\sqrt{g} g^{ij} \partial_j \right)$ on a Riemannian manifold M with or without boundary. For the case of the boundary value problem, the D-N map is defined as $\Lambda(\lambda) : f \to \sum_{i,j} \nu_i g^{ij} \partial_j u \big|_{\partial M}$, where u is a solution to the equation $(H - \lambda)u = 0$ in M, $u = f$ on ∂M, and $\nu = (\nu_1, \cdots, \nu_n)$ is the unit normal on ∂M.

The D-N map $\Lambda(\lambda)$ for all energy λ determines the manifold M and its Riemannian metric $\sum_{i,j} g_{ij} dx^i dx^j$.

For the inverse scattering problem, one considers non-compact manifolds whose non-compact part (end) is asymptotically equal to some standard manifold. We then have:

Given a non-compact manifold $\mathcal{M}$, the knowledge of Riemannian metric of one end $\mathcal{M}_1$ and the associated component of the S-matrix $S_{11}(\lambda)$ for all energies λ determine $\mathcal{M}$.

We will see below more precise statements and various related topics.

5. Input and Output Although the boundary value problem in a bounded domain and the scattering problem on a non-compact space have apparently different nature, they have a feature in common and are mutually related. Let us consider the one-dimensional problem $\left(-d^2/dx^2 + V(x) - \lambda \right) u = 0$ in $(0, \pi)$ with Dirichlet boundary condition $u(0) = 0$. Let $C(x, \lambda)$ and $S(x, \lambda)$ be the solutions of the equation:

$$\left(-\frac{d^2}{dx^2} + V(x) - \lambda \right) y = 0, \quad 0 < x < \pi, \qquad (0.3)$$

satisfying the boundary conditions:

$$C(0, \lambda) = 1, \quad C'(0, \lambda) = 0, \quad S(0, \lambda) = 0, \quad S'(0, \lambda) = 1.$$

Choose $M(\lambda)$ so that

$$y(x, \lambda) = C(x, \lambda) + M(\lambda) S(x, \lambda) \qquad (0.4)$$

satisfies $y(\pi, \lambda) = 0$. This $M(\lambda)$ is the *Weyl function* of the Dirichlet problem for the operator $-d^2/dx^2 + V(x)$. It is also called the *Weyl–Titchmarsh function* or *Weyl's m-function*. Since $M(\lambda) = y'(0, \lambda)$, another definition of Weyl function is the operator:

$$c \to u'(0, \lambda), \qquad (0.5)$$

where $u(x, \lambda)$ is the solution to the Dirichlet problem:

$$\begin{cases} \left(-\dfrac{d^2}{dx^2} + V(x) - \lambda\right)u = 0, & 0 < x < \pi, \\[2mm] u(0) = c, \quad u(\pi) = 0. \end{cases}$$

Let us emphasize here that the Weyl function and the D-N map give equivalent information, as is suggested by (0.4) and (0.5).

The formula (0.4) is a typical representation of the spectral data. One prepares two linearly independent solutions $C(x, \lambda)$, $S(x, \lambda)$ of the Eq. (0.3) at one end point of the domain, so that their linear combination $y(x, \lambda)$ satisfies the boundary condition at the other end point. Then, the coefficient $M(\lambda)$ will contain all the information of the operator $-d^2/dx^2 + V(x)$. Another interpretation is that, sending an input $C(x, \lambda)$ from the boundary $x = 0$, one observes the output $S'(x, \lambda)$ at $x = 0$. The response is represented by $M(\lambda)$. Although (0.4) and (0.5) are simple, they are fundamental formulations of incoming and outgoing data for inverse problems and also used in scattering problems as well as in multi-dimensional problems.

The D-N map and the S-matrix are related in the following way. For the sake of simplicity, let us consider $H = -\sum_{i,j} \frac{1}{\sqrt{g}} \partial_i \left(\sqrt{g} g^{ij} \partial_j\right) + V(x)$ on $\mathbf{R}^n$, where we assume $g_{ij} = \delta_{ij}$ and $V(x) = 0$ outside a bounded open set Ω. Then, one can consider the scattering problem for H on $\mathbf{R}^n$ and also the boundary value problem on Ω. We then have:

The S-matrix $S(\lambda)$ and the D-N map $\Lambda(\lambda)$ determine each other.

Therefore, for compactly supported perturbations, the inverse scattering problem is equivalent to the inverse boundary value problem in a bounded domain.

6. Plan of the Book This book aims at the exposition of basic ideas for solving inverse problems. In Chap. 1, we study the one-dimensional inverse problems on a finite interval and a half-line. It has almost 100 years of history, and we now know a satisfactory answer, which should be used as a basic tool in spectral theory. We summarize an overview of the whole procedure in the one-dimensional problem, not going into the technical details, but paying attention to an intriguing combination of functional analysis and complex function theory.

In Chap. 2, we discuss the multi-dimensional inverse boundary value problem. After briefly overlooking the sources of inverse spectral problems, we introduce an exponentially growing solution for the Schrödinger equation. It has brought a big success both in the inverse boundary value problem and in the inverse scattering problem. We also mention the Carleman estimates bringing a recent rapid progress for multi-dimensional inverse problems.

In Chap. 3, we study the multi-dimensional inverse scattering for Schrödinger operators. The main subject is the multi-dimensional Gel'fand–Levitan theory. In one-dimension, the crucial idea of the Gel'fand–Levitan theory was the transformation to the Volterra integral equation. We analyze this mysterious step by studying its operator theoretical background. Although it still has some formal parts, it explains the trick of inverse scattering from the viewpoint of functional analysis and works as a leading idea for other equations. The new ingredient, $\overline{\partial}$-equation and the characterization result, is also discussed.

In Chap. 4, we discuss the inverse problem on Riemannian manifolds using the boundary control method. As is well known, the Gel'fand–Levitan theory is based on spectral quantities for the stationary wave equation (Helmholtz equation), i.e., in the frequency domain. However, there is another method that has its origin in the one-dimensional spectral theory based on the time-dependent wave equation, i.e., in the time-domain. It is called the boundary control method and developed to be a general method to deal with Riemannian metric. We consider both of compact and non-compact manifolds. In particular, we shall give a solution to the inverse scattering problem on general non-compact manifolds in Theorems 4.12 and 4.14.

Chapter 5 deals with slightly different but closely related problems, which will show new aspects of inverse problems.

In general, solving the inverse problem, especially the inverse scattering problem, requires long arguments, since before going into the inverse problem, one must solve the forward problem. Moreover, the tools are not restricted to the spectral theory. As will be seen in the following arguments, the complex function theory appears very often. In some crucial steps, techniques in the theory of linear partial differential equations (PDEs) play decisive roles. Exposition of these topics and relations to other fields of mathematics is another aim of this book. Due to the limitation of page numbers, we can explain only the main steps of ideas. The proof is given only when it is short and gives the flavor of the essential arguments. Once we know the basic stream of ideas, to follow the details is not a hard problem.

Excellent books have already been published for inverse spectral and scattering problems. For the one-dimensional case, they are summarized in Chap. 1. Multi-dimensional problems, including techniques from partial differential equations and functional analysis, are treated in [84, 104, 147, 166]. The last one aims at numerical computation. The lecture note [35] is close to this book in its spirit with selected and more detailed exposition.

The notation used here is standard. For example, a domain in a manifold M is a connected open subset of M. For a domain Ω in M, $C_0^\infty(\Omega)$ denotes the set of all infinitely differentiable functions with compact support in Ω. We often write $\partial/\partial x_j = \partial_j$ in multi-dimensions.

The author thanks deeply to the referees for their valuable suggestions, which improved significantly this monograph. This work is supported by Grant-in-Aid for Scientific Research (S) 15H05740, (B) 16H0394, Japan Society for the Promotion of Science. The author expresses his gratitude to JSPS.

Kyoto, Japan
June 2020

Hiroshi Isozaki

Contents

Chapter 1
One-Dimensional Inverse Problems

General properties of eigenvalues and eigenfunctions for boundary value problems of second order ordinary differential equations on a line were first studied by Sturm [181, 182] and Liouville [132]. After a long series of special functions introduced in the eighteenth and nineteenth centuries, Weyl [195] gave a unified theory providing a complete proof of the general expansion theorem, which covers both of the cases of discrete and continuous spectrum. It was the dawn of functional analysis, and Weyl used Hellinger's integral [74] for the spectral decomposition of self-adjoint operators. Weyl's work was further developed by Titchmarsh [187] using complex function theory and by Stone [180] and Kodaira [114] on the functional analysis basis. The main concern was the construction of spectral function, the basic idea of which, however, had already been given by Weyl. Nowadays, this key tool is called the Weyl function and plays a central role in the one-dimensional inverse spectral problems. In this chapter, we consider the one-dimensional eigenfunction expansion theory and the ideas of inverse problems.

The first mathematical result for the inverse spectral problem is attributed to Ambarzumian [7]. Around 1950 there appeared a series of seminal papers of one-dimensional inverse problems. The study of the main stream began by the work of Borg [26]. Levinson [129] made a considerable simplification by using analytic function theory. It turned out that in general the knowledge of eigenvalues is not sufficient to determine the potential, and one needs other information of spectral data. It was either the eigenvalues for $-d^2/dx^2 + V(x)$ with different boundary condition or the first order derivative of eigenvectors as mentioned in the introduction. By the *spectral data*, we mean, for the boundary value problem, either two sets of eigenvalues with different boundary conditions or a set of pairs of eigenvalues and the normal derivatives of eigenfunctions at the boundary. For the scattering problem, it is the S-matrix and some information of the discrete spectrum. We have therefore three main ingredients of the spectral problem:

Spectral data,   Weyl function,   Spectral function.

© The Author(s), under exclusive licence to Springer Nature Singapore Pte Ltd. 2020
H. Isozaki, *Inverse Spectral and Scattering Theory*, SpringerBriefs
in Mathematical Physics 38, https://doi.org/10.1007/978-981-15-8199-1_1

Uniqueness of the potential with given Weyl function was proven by Tikhonov [186]. Uniqueness from the spectral function was proven by Marchenko [136], [137] by using the so-called transformation operator. Gel'fand–Levitan [64] developed it to the reconstruction procedure as well as the characterization of the spectral function. The characterization problem for the inverse boundary value problem was studied by Levitan–Gasymov [131], and that for the periodic potential was studied by Marcheko–Ostrovski [139]. For the scattering problem, the results for reconstruction procedure and characterization of the S-matrix are summarized in [138]. The infinite dimensional analytic function approach due to Issacson–Trubowitz [82], Issacson–McKean–Trubowits [83], Dahlberg–Trubowitz [51] is a new viewpoint for the uniqueness, characterization, and isospectral deformation problems. We should also mention the works of Krein [118–121], which turned out to be a new gate to the multi-dimensional problem, and will be discussed in Chap. 4. So many works have been presented for one-dimensional inverse problems that we can quote here only remarkable early papers. One can find more detailed history and a lot of interesting articles in the prefaces and references of monographs [36, 62, 130, 138, 159].

1.1   Functional Analysis and Weyl–Stone–Titchmarsh–Kodaira Theory

1.1.1   Self-Adjoint Extensions

Weyl studied the self-adjointness by using the limit point and limit circle theory. The results are still sharp in the case of one-dimensional differential operators. However, since we are also concerned with the multi-dimensional case, we consider the self-adjointness problem from the viewpoint of functional analysis. For basic notions from functional analysis, see, e. g., [4, 106, 167]. Let $\mathcal{H}$ be a Hilbert space equipped with inner product $(,)$. A densely defined operator A is said to be *symmetric* if $(Au, v) = (u, Av), \forall u, v \in D(A)$. A symmetric operator A is *self-adjoint* if $(w, Au) = (f, u) \forall u \in D(A)$, then $w \in D(A)$ and $Aw = f$. The well-known criterion for the perturbation of self-adjoint operator is the following Kato–Rellich theorem.

Theorem 1.1 *Let A_0 be a self-adjoint operator, and V a closed symmetric operator satisfying $D(A_0) \subset D(V)$, and for some constants $0 < \alpha < 1$, $C > 0$, $\|Vu\| \leq \alpha \|A_0 u\| + C \|u\|$ holds for any $u \in D(A)$. Then, $A_0 + V$ with domain $D(A_0 + V) = D(A_0)$ is self-adjoint.*

Why the self-adjointness is important? For a symmetric operator A, the self-adjointness of A is equivalent to the surjectivity of $A \pm i : D(A) \to \mathcal{H}$ or the injectivity of $A^* \pm i : D(A) \to \mathcal{H}$. If A is (an extension of) a differential operator, this means the solvability of the differential equation $(A - z)u = f$ or the uniqueness for $(A^* \pm i)u = 0$ under the prescribed boundary condition. This is a highly non-trivial question especially when A is defined on a non-compact domain, or the coefficient of A is singular, or A is a partial differential operator.

In realizing a differential operator as a self-adjoint operator, as is seen from the above definition, one needs to consider a weak form of the differential equation by forming inner products. This leads us to the Sobolev space. Let I be an open interval in $\mathbf{R}^1$. For $f, g \in L^2(I)$, putting $(f, g) = \int_I f(x)\overline{g(x)}dx$, we say that g is a *weak derivative* of the n-th order of f, if $(f, \varphi^{(n)}) = (-1)^n(g, \varphi)$, $\forall \varphi \in C_0^\infty(I)$, holds. We denote the n-th order weak derivative of f by $f^{(n)}$ or $d^n f/dx^n$. For a non-negative integer m, the *Sobolev space* $H^m(I)$ is defined to be the set of all $f \in L^2(I)$ whose all weak derivatives up to order m are in $L^2(I)$. Equipped with the inner product $(f, g)_{H^m(I)} = \sum_{n=0}^m (f^{(n)}, g^{(n)})$, $H^m(I)$ is a Hilbert space. Let $H_0^m(I)$ be the closure of $C_0^\infty(I)$ by H^m-norm, which is the set of $u \in H^m(I)$ such that the derivatives of u up to order $m - 1$ vanish at the boundary of I.

Let $L_{loc}^1(I)$ be the set of measurable functions $f(x)$ on I such that $\int_K |f(x)|dx < \infty$ for any compact set $K \subset I$. We put $\langle f, \varphi \rangle = \int_I f(x)\varphi(x)dx$ for $f \in L_{loc}^1(I)$ and $\varphi \in C_0^\infty(I)$. Let $a(x), b(x) \in C^\infty(I)$ and $g(x) \in L_{loc}^1(I)$. Then, $u(x) \in L_{loc}^1(I)$ is said to be a *weak solution* of the differential equation

$$u'' + a(x)u' + b(x)u = g(x), \tag{1.1}$$

if for any $\varphi \in C_0^\infty(I)$

$$\langle u, \varphi'' - (a\varphi)' + b\varphi \rangle = \langle g, \varphi \rangle$$

holds. The following theorem shows that the weak solution of the Eq. (1.1) is also a solution of (1.1) in the usual sense .

Theorem 1.2 *Let $u(x) \in L_{loc}^1(I)$ be a weak solution of (1.1). Then, $u(x) \in C^1(I)$, $u'(x)$ is absolutely continuous on any compact subinterval in I, and (1.1) holds almost everywhere on I. For $I = (a, b)$ with finite a, take any $a < c < b$ and put $I_c = (a, c)$. If $u(x), g(x) \in L^1(I_c)$ and $a(x), a'(x), b(x)$ are bounded on I_c, $u(x), u'(x)$ are continuously extended to a.*

Now, we consider the Schrödinger operators

$$H_0 = -\frac{d^2}{dx^2}, \quad H = H_0 + q(x), \quad \text{on} \quad I = (0, \ell),$$

where $\ell > 0$. We allow the case $\ell = \infty$. Assume that $q(x) \in L^2(I)$ and is real-valued. The domain of H is defined as follows. When $\ell < \infty$, $D(H)$ is the set of $u \in H^2(I)$ such that

$$a_0 u'(0) - b_0 u(0) = 0, \quad a_\ell u'(\ell) + b_\ell u(\ell) = 0, \tag{1.2}$$

where $a_0, b_0, a_\ell, b_\ell \in \mathbf{R}$ satisfy

$$a_0^2 + b_0^2 = a_\ell^2 + b_\ell^2 = 1.$$

When $\ell = \infty$, only the condition at 0 is assumed in (1.2). The self-adjointness of H_0 follows from Theorem 1.2.[1] Theorem 1.1 then implies the self-adjointness of H.[2]

Theorem 1.3 *The operator H is self-adjoint in $L^2(I)$.*

In the application to scattering theory in one-dimension, the potential $q(x)$ may be singular at $x = 0$. When we assume the Dirichlet boundary condition at $x = 0$, the above theorem allows the singular potential like $q(x) = O(x^{-\delta})$, $\delta < 3/2$. However the theory of quadratic forms makes it possible to deal with the stronger singularity $q(x) = O(x^{-\delta})$, $\delta < 2$.

Let $a(\cdot, \cdot)$ be a sesquilinear form (often called quadratic form) on a Hilbert space $\mathscr{H}$ with domain D, a dense subspace of $\mathscr{H}$, i.e. a mapping : $D \times D \to \mathbf{C}$ satisfying

$$a(\lambda u + \mu v, w) = \lambda a(u, w) + \mu a(v, w), \quad u, v, w \in D, \quad \lambda, \mu \in \mathbf{C},$$

$$\overline{a(u, v)} = a(v, u), \quad u, v \in D.$$

Assume that a is positive definite, i.e. there exists a constant $C > 0$ such that

$$a(u, u) \geq C\|u\|^2, \quad u \in D.$$

Define a new norm $\|u\|_a$ by $\|u\|_a = \sqrt{a(u, u)}$, $u \in D$. The quadratic form a is said to be closed if D is complete with respect to this norm $\|u\|_a$.

Theorem 1.4 *Let $a(\ ,\ )$ be a positive definite closed form with domain D. Then, there exists a unique self-adjoint operator A such that $D = D(A^{1/2})$, and*

$$a(u, v) = (Au, v), \quad u \in D(A), \quad v \in D$$

Here let us recall that for any self-adjoint operator A, there exists a family of orthogonal projections $\{E(\lambda)\}_{\lambda \in \mathbf{R}}$, called the spectral decomposition of A, which is right-continuous and monotone increasing in λ, such that A is written as an operator-valued Stieltjes integral $A = \int_{-\infty}^{\infty} \lambda \, dE(\lambda)$. If A is non-negative, i.e.

---

[1] To show the self-adjointness of H_0 is a good exercise for the 1-dim. Sobolev space. The main problem is the regularity of $u \in D(H_0^*)$. For the case $I = (-\infty, \infty)$, a simple method is to pass to the Fourier transform. For the case $(0, \infty)$, one first shows that $u \in D(H_0^*)$ is in $H^1((-\infty, \infty))$. Then one may cut off $u \in D(H_0^*)$ into $u = u_0 + u_1$, where u_1 is supported in $(1, \infty)$ hence $u_1 \in H^2((-\infty, \infty))$, and u_0 is supported in $(0, 2)$, hence shown to be in $H^2((0, 2))$ by integration by parts.

[2] We also use the inequality $|u(x)| \leq \epsilon\|u\|_{H^1(I)} + C_\epsilon\|u\|_{L^2(I)}, \quad \forall \epsilon > 0, \quad x \in I.$

$(Au, u) \geq 0, u \in D(A)$, then $E(\lambda) = 0$ for $\lambda < 0$, and $A^{1/2}$ is defined by $A^{1/2} = \int_0^\infty \lambda^{1/2} dE(\lambda)$ with domain $D(A^{1/2}) = \{u \in \mathscr{H} \, ; \, \int_0^\infty \lambda d(E(\lambda)u, u) < \infty\}$.

Perturbations of this operator A are defined by quadratic forms.

Theorem 1.5 *Let A be a non-negative self-adjoint operator, b a quadratic form with domain $D(A^{1/2})$ satisfying*

$$b(u, v) = (A^{1/2}u, A^{1/2}v) + r(u, v), \quad u, v \in D(A^{1/2}),$$

and there exist constants $0 < \delta < 1$, $C > 0$ such that

$$|r(u, u)| \leq \delta \|A^{1/2}u\|^2 + C\|u\|^2, \quad u \in D(A^{1/2}).$$

Then, there exists a unique self-adjoint operator B such that $D(B) \subset D(A^{1/2})$, and

$$(u, Bv) = b(u, v), \quad u \in D(A^{1/2}), \quad v \in D(B),$$

$$B \geq -C, \quad D((B + C)^{1/2}) = D(A^{1/2}).$$

As an example, consider the quadratic form $q_{D,0}$ on $L^2(0, \infty)$ defined by $q_{D,0}(u, v) = (u', v')$, with domain $D(q_{D,0}) = H_0^1(I)$, $I = (0, \infty)$. Then, the associated self-adjoint operator $H_{D,0}$ is given by

$$H_{D,0} = -\frac{d^2}{dx^2}, \quad D(H_{D,0}) = H^2(I) \cap H_0^1(I), \quad D(H_{D,0}^{1/2}) = H_0^1(I).$$

As a perturbation, we can allow potentials $q(x)$ satisfying[3]

$$\int_0^\infty x|q(x)|dx < \infty. \tag{1.3}$$

For a self-adjoint operator A in a Hilbert space $\mathscr{H}$, the *resolvent set* of A, denoted by $\rho(A)$, is defined as the set of $z \in \mathbf{C}$ such that $A - z : D(A) \to \mathscr{H}$ is one to one and onto, hence $(A - z)^{-1}$ exists. The *spectrum* of A is defined by $\sigma(A) = \mathbf{C} \setminus \rho(A)$. It is a closed set in $\mathbf{R}$, and decomposed as $\sigma(A) = \sigma_d(A) \cup \sigma_e(A)$, where $\sigma_d(A)$, called the *discrete spectrum* of A, is the set of isolated eigenvalues of A with finite multiplicities, and $\sigma_e(A) = \sigma(A) \setminus \sigma_d(A)$ is the *essential spectrum* of A, which are eigenvalues of infinite multiplicities or accumulation points of the spectrum.

Using the spectral decomposition $\{E(\lambda)\}_{\lambda \in \mathbf{R}}$ of A, $\mathscr{H}$ is decomposed as an orthogonal sum: $\mathscr{H} = \mathscr{H}_{ac}(A) \oplus \mathscr{H}_{sc}(A) \oplus \mathscr{H}_{pp}(A)$, where $(E(\lambda)u, u)$ is absolutely continuous with respect to the Lebesgue measure for $u \in \mathscr{H}_{ac}(A)$, singular continuous for $u \in \mathscr{H}_{sc}(A)$ and an increasing step function for $u \in \mathscr{H}_{pp}(A)$.

---

[3]We use the inequality $|u(x)| \leq \sqrt{x}\|u'\|_{L^2((0,1))}, \quad 0 < x < 1, \quad u \in H_0^1(I).$

Hence $\mathscr{H}_{pp}(A)$ is the closure of the linear combinations of eigenvectors of A. The *absolutely continuous spectrum* $\sigma_{ac}(A)$ and *singular continuous spectrum* $\sigma_{sc}(A)$ are defined as the spectrum of A restricted to $\mathscr{H}_{ac}(A)$ and $\mathscr{H}_{sc}(A)$, respectively. The space $\mathscr{H}_{cont}(A) = \mathscr{H}_{ac}(A) \oplus \mathscr{H}_{sc}(A)$ is called the continuous spectral subspace. The spectrum of A restricted to $\mathscr{H}_{cont}(A)$ is called the *continuous spectrum*, and denoted by $\sigma_c(A)$.

Now, we consider the Schrödinger operator $H = -d^2/dx^2 + q(x)$ defined above.

Theorem 1.6

(1) For a bounded interval I, $\mathscr{H} = \mathscr{H}_{pp}(H)$ and $\sigma(H) = \sigma_d(H)$.

(2) For the interval $I = (0, \infty)$, $\sigma_d(H) \subset (-\infty, 0)$, and $\sigma_e(H) = [0, \infty)$.

In fact, we have seen that $(H - z)^{-1}$ is a bounded operator from $L^2(I)$ to $H^1(I)$. This implies that $(H - z)^{-1}$ is a compact operator for the case $I = (0, \pi)$ by virtue of Rellich's selection theorem.[4] The theory of compact self-adjoint operators then implies (1). To prove (2), we note that $(H - z)^{-1} - (H_0 - z)^{-1}$ is compact, where $H_0 = H$ for the case $q = 0$. Weyl's theorem on the perturbation of essential spectrum then yields $\sigma_e(H) = \sigma_e(H_0) = \sigma(H_0) = [0, \infty)$. Letting $N(q)$ be the number of negative eigenvalues of $H = -d^2/dx^2 + q$ and $q_-(x)$ the negative part of $q(x)$, one can show

$$N(q) \leq \int_0^\infty x|q_-(x)|dx.$$

1.1.2   Weyl Function

Let $I \subset \mathbf{R}$ be an interval and a one of its finite end point. Letting h_a be a real constant, we define the boundary operators B_D and B_R by

$$B_D y = y(a), \quad B_R y = y'(a) + h_a y(a), \quad y \in H^1(I).$$

We begin with the case $I = (0, \pi)$, and consider the equation

$$-y'' + q(x)y = \lambda y \quad \text{on} \quad I \tag{1.4}$$

with parameter $\lambda \in \mathbf{C}$. Assume that $q(x) \in L^2(I)$. We consider two boundary conditions

$$(D) \quad B_D y = 0 \quad \text{at} \quad x = 0, \quad \text{or} \quad (R) \quad B_R y = 0 \quad \text{at} \quad x = 0.$$

---

[4]Let I be a finite interval. Then, the imbedding operator $H^1(I) \ni u \to u \in L^2(I)$ is compact.

Let $\varphi_D(x, \lambda)$, $\varphi_R(x, \lambda)$ be the solutions to (1.4) satisfying at $x = 0$

$$B_D\dot{\varphi_D} = 0, \quad B_R\varphi_D = 1, \quad B_R\varphi_R = 0, \quad B_D\varphi_R = 1.$$

Note that φ_D and φ_R are linearly independent. At the other end point $x = \pi$, we fix any boundary condition of the type (D) or (R), whose boundary operator is denoted by $\widetilde{B}^\pi$. The operators $-d^2/dx^2 + q(x)$ are then self-adjoint, which we denote by H_D and H_R. Let $\psi_D(x, \lambda)$, $\psi_R(x, \lambda)$ be the solutions to (1.4) satisfying

$$\begin{aligned}
B_D\psi_D = 1 \quad \text{at} \quad x = 0, \quad \widetilde{B}^\pi\psi_D = 0 \quad \text{at} \quad x = \pi, \\
B_R\psi_R = 1 \quad \text{at} \quad x = 0, \quad \widetilde{B}^\pi\psi_R = 0 \quad \text{at} \quad x = \pi.
\end{aligned} \tag{1.5}$$

By the self-adjointness of H_D and H_R, they exist for λ outside of the spectrum.

Definition 1.1 *Weyl functions* $M_D(\lambda)$, $M_R(\lambda)$ are defined by the formula

$$\psi_D(x, \lambda) = \varphi_R(x, \lambda) + M_D(\lambda)\varphi_D(x, \lambda),$$

$$\psi_R(x, \lambda) = \varphi_D(x, \lambda) + M_R(\lambda)\varphi_R(x, \lambda).$$

Note that

$$M_D(\lambda) = B_R\psi_D(0, \lambda), \quad M_R(\lambda) = B_D\psi_R(0, \lambda). \tag{1.6}$$

Therefore, $M_D(\lambda)$ is meromorphic in $\mathbf{C}$ with poles at $\sigma(H_D)$, and $M_R(\lambda)$ is meromorphic in $\mathbf{C}$ with poles at $\sigma(H_R)$. Let $y(x, \lambda)$ be a solution to (1.4) satisfying $\widetilde{B}^\pi y = 0$. The *Dirichlet-to-Robin map (D-R map)* $\Lambda_D(\lambda)$ and the *Robin-to-Dirichlet map (R-D map)* $\Lambda_R(\lambda)$ are defined by

$$\Lambda_D(\lambda) : y(0, \lambda) \rightarrow y'(0, \lambda) + h_0 y(0, \lambda),$$

$$\Lambda_R(\lambda) : y'(0, \lambda) + h_0 y(0, \lambda) \rightarrow y(0, \lambda).$$

They are the operators of multiplication by $M_D(\lambda)$ and $M_R(\lambda)$. Therefore,

$$M_D(\lambda)M_R(\lambda) = 1. \tag{1.7}$$

First let us consider $M_R(\lambda)$. Let $\lambda_0^{(R)} < \lambda_1^{(R)} < \lambda_2^{(R)} < \cdots$ be the eigenvalues of H_R and $\varphi_n^{(R)}(x)$ the associated orthonormal system of eigenvectors. Then $\psi_R(x, \lambda)$

is written as[5]

$$\psi_R(x, \lambda) = \sum_{n=0}^{\infty} \frac{\varphi_n^{(R)}(0)}{\lambda - \lambda_n^{(R)}} \varphi_n^{(R)}(x). \tag{1.8}$$

Since $\varphi_n^{(R)}(x) = \varphi_R(x, \lambda_n^{(R)})/\|\varphi_R(\cdot, \lambda_n^{(R)})\|$, we have the following representation of the Weyl function $M_R(\lambda)$.

Lemma 1.1 *Letting $\{\lambda_n^{(R)}\}_{n=0}^{\infty}$ and $\{\varphi_n^{(R)}(x)\}_{n=0}^{\infty}$ be the eigenvalues and orthonormal system of eigenvectors of H_R, we have*

$$M_R(\lambda) = \sum_{n=0}^{\infty} \frac{|\varphi_n^{(R)}(0)|^2}{\lambda - \lambda_n^{(R)}} = \sum_{n=0}^{\infty} \frac{1}{(\lambda - \lambda_n^{(R)})\|\varphi^{(R)}(\cdot, \lambda_n^{(R)})\|_{L^2(I)}^2} \tag{1.9}$$

Therefore, the poles of $M_R(\lambda)$ are the eigenvalues $\lambda_n^{(R)}$ and the residue is $\varphi_n^{(R)}(0)^2$. Since the sign of eigenfunctions does not matter, we can recover the eigenvalues and $\varphi_n^{(R)}(0)$ from $M_R(\lambda)$.

Next we consider $M_D(\lambda)$. By a similar computation, we have

$$\psi_D(x, \lambda) = \sum_{n=0}^{\infty} \frac{\varphi'^{(D)}_n(0)}{\lambda_n^{(D)} - \lambda} \varphi_n^{(D)}(x)$$

with obvious notation. We then have, formally, $M_D(\lambda) = \sum_{n=0}^{\infty} \frac{|\varphi'^{(D)}_n(0)|^2}{\lambda_n^{(D)} - \lambda}$. However, it does not converge.[6] The following lemma shows that $M_D(\lambda)$ still contains the information of spectrum.

Lemma 1.2 *Let $\{\lambda_n^{(D)}\}_{n=0}^{\infty}$ be the eigenvalues of H_D and $\{\varphi_n^{(D)}(x)\}_{n=0}^{\infty}$ the associated normalized eigenvectors. Then, the Weyl function $M_D(\lambda)$ is meromorphic in $\mathbf{C}$ with poles at eigenvalues $\lambda_n^{(D)}$, and $\varphi'^{(D)}_n(0)$ can be recovered from its residue.[7]*

---

[5]Take $\chi(x) \in C^{\infty}(\mathbf{R})$ such that $B_R\chi(0) = 1$ and $\chi(x) = 0$ for $x > \pi/2$. Letting $f = -\chi'' + q\chi - \lambda\chi$, we have $\psi_R(x, \lambda) = \chi(x) - (H_R - \lambda)^{-1} f = \chi(x) - \sum_{n=0}^{\infty} \frac{(f, \varphi_n^{(R)})}{\lambda_n^{(R)} - \lambda} \varphi_n^{(R)}(x)$. Integration by parts yields $(f, \varphi_n^{(R)}) = \varphi_n^{(R)}(0) - (\lambda - \lambda_n^{(R)})(\chi, \varphi_n^{(R)})$, which proves (1.8).

[6]Letting $\varphi_n(x)$ be either $\varphi_n^{(D)}(x)$ or $\varphi_n^{(R)}(x)$, one can show that $\varphi_n(x) = O(1)$, $\varphi_n'(x) = O(n)$ as $n \to \infty$. Letting λ_n be either $\lambda_n^{(D)}$ or $\lambda_n^{(R)}$, we have $\lambda_n \sim n^2$.

[7]By $\psi_D(x, \lambda) = \chi - (H_D - \lambda)^{-1} f$, $\psi_D(x, \lambda)$ is meromorphic with respect to λ with poles at $\sigma_d(H_D)$. Integrating along a contour around $\lambda_n^{(D)}$ and applying B_R, we get Lemma 1.2.

Let us consider the case $I = (0, \infty)$ assuming (1.3). We convert the differential equation

$$- y'' + q(x)y = k^2 y, \quad k \in \mathbf{C} \tag{1.10}$$

into an integral equation. If $\phi \in C^1([0, \infty))$ is a solution to (1.10) such that $\phi(0) = 0, \phi'(0) = 1$, then ϕ satisfies the integral equation

$$\phi(x, k) = \frac{\sin kx}{k} + \int_0^x \frac{\sin k(x - t)}{k} q(t)\phi(t, k)dt, \quad k \in \mathbf{C}. \tag{1.11}$$

Conversely, if $\phi \in C([0, \infty))$ is a solution to (1.11) satisfying $|\phi(x, k)| \leq Cx$, where C is a constant independent of $0 < x < 1$, then ϕ satisfies (1.10), moreover, $\phi \in C^1([0, \infty))$ and

$$\phi(0, k) = 0, \quad \phi'(0, k) = 1, \quad \phi(x, k) = \phi(x, -k).$$

Note that $\phi(x, k)$ is an entire function of $k \in \mathbf{C}$.

At $x = \infty$, the natural boundary condition is that $y \in L^2((1, \infty))$. Therefore, we take $k \in \overline{\mathbf{C}_+}$[8] and seek a solution to (1.10) which behaves like e^{ikx} near $x = \infty$. It leads us to the integral equation

$$f(x, k) = e^{ikx} - \frac{1}{k} \int_x^\infty \sin k(x - t)q(t)f(t, \lambda) \, dt, \quad k \in \overline{\mathbf{C}_+}, \tag{1.12}$$

the solution of which is called *Jost solution*. The condition (1.3) is appropriate to solve (1.11) and (1.12) by iteration.

Lemma 1.3 *Assume (1.3). Then, for any fixed $k \in \overline{\mathbf{C}_+}$, $f(x, k)$ is bounded continuous on $[0, \infty)$, moreover $\lim_{x \to 0} xf'(x, k) \to 0$.[9] As a function of k, $f(x, k)$ is analytic on $\mathbf{C}_+$, continuous on $\overline{\mathbf{C}_+}$.*

We define the Wronskian by

$$[f(x), g(x)] = f(x)g'(x) - f'(x)g(x),$$

and put

$$w(k) = [f(x, k), \phi(x, k)] = f(0, k), \quad k \in \overline{\mathbf{C}_+}. \tag{1.13}$$

---

[8]Here, $\mathbf{C}_+ = \{k \in \mathbf{C} ; , \operatorname{Im} k > 0\}$, $\overline{\mathbf{C}_+} = \{k \in \mathbf{C} ; , \operatorname{Im} k \geq 0\}$.
[9]Use $xf' = ikxe^{ikx} - \int_0^\infty a_x(t)tV(t)f(t, k)dt$, where $a_x(t) = \cos k(x - t)\chi_x(t)\frac{x}{t}$, and $\chi_x(t)$ is the characteristic function of (x, ∞). Noting $|a_x(t)| \leq 1$, apply Lebesgue's convergence theorem.

If $w(k) = 0$, $f(x, k) = c(k)\phi(x, k)$ for a constant $c(k)$. If $k \notin \mathbf{R}$, this means that k^2 is an eigenvalue of H. If $k \in \mathbf{R} \setminus \{0\}$, $\phi(x, k) = c(k)^{-1} f(x, k)$ is real-valued. Observing its behavior as $x \to \infty$, we arrive at a contradiction. This proves the following lemma.

Lemma 1.4

(1) $w(k)$ is analytic on $\mathbf{C}_+$, and continuous on $\overline{\mathbf{C}_+} \setminus \{0\}$.
(2) $w(k) \neq 0$ for $k \in \mathbf{R} \setminus \{0\}$.
(3) $w(k)$ vanishes on $\mathbf{C}_+$ if and only if $k = ik_j$, where $-k_j^2 \in \sigma_d(H)$.

The following formula plays an important role in the eigenfunction expansion:[10]

$$\phi(x, k) = \frac{1}{2ik}\Big(f(x, k)w(-k) - f(x, -k)w(k)\Big), \quad k > 0. \tag{1.14}$$

We take the branch of $\sqrt{\lambda}$ so that $\mathrm{Im}\sqrt{\lambda} > 0$ for $\lambda \in \mathbf{C} \setminus [0, \infty)$. Hence $\sqrt{\lambda \pm i\epsilon} \to \pm\sqrt{\lambda}$ for $\lambda > 0$ as $\epsilon \to 0$. We put

$$\varphi(x, \lambda) = \phi(x, \sqrt{\lambda}), \quad \psi(x, \lambda) = f(x, \sqrt{\lambda}),$$

which correspond to $\varphi_D(x, \lambda)$ and $\psi_D(x, \lambda)$ in the case $I = (0, \pi)$. For the case $I = (0, \infty)$, we use information of the Jost function instead of Weyl function.[11]

1.1.3   Eigenfunction Expansion

Let H be any of the Schrödinger operators introduced in Sect. 1.1.2. We omit the sub (super) script D or R of all relevant operators.

Let $I = (0, \ell)$, where $\ell = \pi$ or ∞. For $\lambda \in \mathbf{C}$ and $f \in L^2(I)$ of compact support, we define the operator $\mathscr{F}(\lambda)$ by

$$\mathscr{F}(\lambda)f = \int_I \varphi(x, \lambda) f(x) dx. \tag{1.15}$$

---

[10]Putting $g(x, k) = f(x, k)w(-k) - f(x, -k)w(k)$, by (1.13), $g(x, k) = [f(x, -k), f(x, k)]\phi(x, k)$. Since $f(x, k) \sim e^{ikx}$ as $x \to \infty$, we have $[f(x, -k), f(x, k)] = 2ik$, which proves (1.14).

[11]If $q(x) \in L^2((0, 1))$, the Weyl function $M_D(\lambda)$ is defined by using Dirichlet and Robin boundary conditions at $x = 0$, and is written as $M_D(\lambda) = \frac{f'(0, \sqrt{\lambda})}{f(0, \sqrt{\lambda})} + h_0$, $\sqrt{\lambda} \in \overline{\mathbf{C}_+}$. This suggests that in the case of $(0, \infty)$, the Jost solution plays a central role.

The eigenfunction expansion theorem says that there is a monotone increasing function $\rho(\lambda)$, called the *spectral function*, by which any $f \in L^2(I)$ can be expanded as

$$f(x) = \int_{-\infty}^{\infty} \varphi(x, \lambda) \mathscr{F}(\lambda) f \, d\rho(\lambda). \tag{1.16}$$

Recall that $\varphi(x, \lambda)$ and $\psi(x, \lambda)$ are solutions of the equation $-y'' + q(x)y = \lambda y$ satisfying the prescribed boundary conditions at $x = 0$ and $x = \pi$ or ∞. Letting

$$W(\lambda) = w(\sqrt{\lambda}) = [\psi, \varphi] = \psi(x, \lambda)\varphi'(x, \lambda) - \psi'(x, \lambda)\varphi(x, \lambda),$$

we define the *Green function* by

$$R(x, y; \lambda) = \frac{1}{W(\lambda)} \begin{cases} \psi(x, \lambda)\varphi(y, \lambda), & 0 < y < x < \ell, \\ \varphi(x, \lambda)\psi(y, \lambda), & 0 < x < y < \ell. \end{cases}$$

It is the integral kernel of the resolvent of H:

$$R(\lambda)f = (H - \lambda)^{-1} f = \int_I R(x, y; \lambda) f(y) dy, \quad \lambda \notin \sigma(H). \tag{1.17}$$

Recall *Stone's formula* (see, e.g., [167], Vol 1, p. 237).

Theorem 1.7 *For a self-adjoint operator* $A = \int_{-\infty}^{\infty} \lambda \, dE(\lambda)$,

$$\left(\frac{1}{2} \Big[E((a, b)) + E([a, b]) \Big] f, g \right) = \lim_{\epsilon \downarrow 0} \frac{1}{2\pi i} \int_a^b \Big((R_A(\lambda + i\epsilon) - R_A(\lambda - i\epsilon)) f, g \Big) d\lambda$$

holds for any $a < b$, *where* $R_A(z) = (A - z)^{-1}$.

It implies that if $\eta \in \sigma_d(A)$, the associated eigenprojection P_η is, for small $\epsilon > 0$,

$$P_\eta = -\frac{1}{2\pi i} \int_{|\eta - \lambda| = \epsilon} R_A(\lambda) d\lambda. \tag{1.18}$$

For the case $I = (0, \pi)$, the *spectral function* is given by[12]

$$\rho(\lambda) = \int_{-\infty}^{\lambda} \sum_{n=0}^{\infty} \frac{1}{\alpha_n} \delta(\lambda - \lambda_n) \, d\lambda = \sum_{\lambda_n \leq \lambda} \frac{1}{\alpha_n}, \tag{1.19}$$

---

[12]$\delta(\lambda - t)d\lambda$ is the Dirac measure supported at t.

where α_n is the *normalizing constant* defined by

$$\alpha_n = \begin{cases} \int_0^\pi \varphi_R(x, \lambda_n^{(R)})^2 dx = \dfrac{1}{|\varphi_n^{(R)}(0)|^2}, & \text{for} \quad H_R, \\[4mm] \int_0^\pi \varphi_D(x, \lambda_n^{(D)})^2 dx = \dfrac{1}{|\varphi_n'^{(D)}(0)|^2}, & \text{for} \quad H_D. \end{cases} \tag{1.20}$$

In fact, for H_R, $W(\lambda) = -1$, hence by (1.1), $R(\lambda)f = -M(\lambda)\varphi(x, \lambda)\mathscr{F}(\lambda)f +$ an entire function. By (1.18), (1.17), and (1.9), we have for any $f, g \in L^2(I)$

$$(f, g) = \sum_{n=0}^\infty \frac{-1}{2\pi i} \int_{|\lambda_n - \lambda| = \epsilon} (R(\lambda)f, g)d\lambda = \sum_{n=0}^\infty \frac{1}{\alpha_n} \mathscr{F}(\lambda_n)f \overline{\mathscr{F}(\lambda_n)g}.$$

Then, we have for any $f \in L^2(I)$

$$f(x) = \sum_{n=0}^\infty \frac{1}{\alpha_n} \varphi(x, \lambda_n)\mathscr{F}(\lambda_n)f,$$

which implies (1.16). The Dirichlet case is proved similarly using $W(\lambda) = 1$.

For the case $I = (0, \infty)$, there is no eigenvalue in $(0, \infty)$. A fact peculiar to the one-dimension is that 0 is not an eigenvalue of H.[13] Let $-k_j^2$, $1 \le j \le n$, be the negative eigenvalues of H, and P_j the associated eigenprojection. Since ik_j is a simple zero of $w(k)$,[14] we have by using (1.18)

$$E((-\infty, 0]) = \sum_{j=1}^n P_j, \quad P_j f = \frac{1}{\gamma_j} \varphi(x, -k_j^2)\mathscr{F}(-k_j^2)f,$$

$$\gamma_j = \int_0^\infty \varphi(x, -k_j^2)^2 dx, \quad j = 1, \cdots, n.$$

Take $0 < a < b < \infty$ arbitrarily. Then, by Stone's formula,

$$E((a, b)) = \lim_{\epsilon \to 0} \frac{1}{2\pi i} \int_a^b \Big(R(\lambda + i\epsilon) - R(\lambda - i\epsilon) \Big)d\lambda. \tag{1.21}$$

---

[13]Let $y(x)$ be an eigenfunction associated with eigenvalue 0. Then $[y(x), \varphi(x, 0)] = 0$, hence $y(x) = c\varphi(x, 0)$. However, $\varphi(x, 0) \notin L^2((0, \infty))$.

[14]For this and (1.22), see [138], Lemma 3.1.6.

By virtue of (1.14), the integral kernel of the right-hand side converges to

$$\frac{1}{2\pi i}\Big(R(x, y; \lambda + i0) - R(x, y; \lambda - i0)\Big) = \frac{\sqrt{\lambda}}{\pi |W(\lambda)|^2}\varphi(x, \lambda)\varphi(y, \lambda).$$

With this in mind, we define the *spectral function* of H by

$$\rho(\lambda) = \begin{cases} 0, & \lambda < -k_1^2 \\[2mm] \dfrac{1}{\gamma_1} + \cdots + \dfrac{1}{\gamma_j}, & -k_j^2 \le \lambda < -k_{j+1}^2, \\[3mm] \dfrac{1}{\gamma_1} + \cdots + \dfrac{1}{\gamma_n} + \displaystyle\int_0^\lambda \frac{\sqrt{t}}{\pi |W(t)|^2}dt, & 0 \le \lambda, \end{cases}$$

where $k_{n+1} = 0$. It is known that there exist constants $C, \delta > 0$ such that

$$\left|\frac{k}{f(0, k)}\right| \le C, \quad k \in \overline{\mathbf{C}_+}, \quad |k| \le \delta. \tag{1.22}$$

Then, for any $f \in L^2((0, \infty))$ with compact support, we have

$$(E(S)f, f) = \int_S |\mathscr{F}(\lambda)f|^2 \rho'(\lambda)d\lambda, \tag{1.23}$$

where $S = (a, b), 0 < a < b < \infty$. Letting $a \to 0, b \to \infty$, (1.23) extends to any $f \in L^2((0, \infty))$ and $S = (0, \infty)$. This proves

$$E([0, \infty))L^2((0, \infty)) = \mathscr{H}_{ac}(H),$$

and the following orthogonal decomposition

$$L^2((0, \infty)) = \mathscr{H}_{ac}(H) \oplus \mathscr{H}_{pp}(H).$$

We have also obtained Parseval's formula

$$\|f\|^2 = \int_0^\infty |\mathscr{F}(\lambda)f|^2 \rho'(\lambda)d\lambda + \sum_{j=1}^n \frac{1}{\gamma_j}|\mathscr{F}(-k_j^2)f|^2.$$

We now put

$$\mathbf{H} = L^2((0, \infty)), \quad \widehat{\mathbf{H}} = L^2(\sigma(H); d\rho),$$

the latter being the set of functions $\widehat{f}(\lambda)$ defined for $\lambda \in \sigma(H)$ such that

$$\|\widehat{f}\|_{\widehat{\mathbf{H}}}^2 = \int_{-\infty}^{\infty} |\widehat{f}(\lambda)|^2 d\rho(\lambda) = \int_0^{\infty} |\widehat{f}(\lambda)|^2 \rho'(\lambda) d\lambda + \sum_{j=1}^n \frac{1}{\gamma_j} |\widehat{f}(-k_j^2)|^2 < \infty.$$

We have now arrived at the following eigenfunction expansion theorem.

Theorem 1.8

(1) *The operator $\mathscr{F}$ defined by $(\mathscr{F}f)(\lambda) = \mathscr{F}(\lambda)f$ for compactly supported $f \in$* **H** *is uniquely extended to a unitary operator from* **H** *to* $\widehat{\mathbf{H}}$, *which is denoted by $\mathscr{F}$ again.*

(2) *For $f \in D(H)$, we have*

$$(\mathscr{F}Hf)(\lambda) = \lambda(\mathscr{F}f)(\lambda).$$

(3) *For any $f \in$* **H** *and $N > 0$, $\int_{-\infty}^N \varphi(x,\lambda)(\mathscr{F}f)(\lambda)d\rho(\lambda) \in$* **H**. *Moreover,*

$$f = \mathrm{s} - \lim_{N \to \infty} \int_{-\infty}^N \varphi(x,\lambda)(\mathscr{F}f)(\lambda)d\rho(\lambda) \quad in \quad \mathbf{H}.$$

For $H = -d^2/dx^2$ with Dirichlet or Neumann condition at $x = 0$, $\mathscr{F}$ is the Fourier sine or cosine transformation. Therefore, $\mathscr{F}$ is often called the *generalized Fourier transformation*. Any element in the absolutely continuous subspace $\mathscr{H}_{ac}(H)$ is expanded by $\varphi(x,\lambda)$, Therefore, $\varphi(x,\lambda)$ is often called the generalized plane wave.

1.2 Spectral Data on the Finite Interval

Let us study the relations between eigenvalues $\{\lambda_n\}$, normalizing constants $\{\alpha_n\}$, Weyl function $M(\lambda)$, and spectral function $\rho(\lambda)$. Almost all results in the one-dimensional inverse problem do not depend on the boundary condition. For the sake of simplicity, we mainly consider the case $H = H_R$:

$$Hy = -y'' + q(x)y = \lambda y, \quad 0 < x < \pi, \tag{1.24}$$

$$B_R y = y'(0) + h_0 y(0) = 0, \quad \widetilde{B}_R y = y'(\pi) + h_\pi y(\pi) = 0.$$

Let $\varphi_R(x,\lambda)$, $\widetilde{\varphi}_R(x,\lambda)$ be a fundamental system of solutions satisfying (1.24) and the following boundary conditions:

$$\varphi_R(0,\lambda) = 1, \quad \varphi_R'(0,\lambda) = -h_0, \quad \text{hence} \quad B_R\varphi_R = 0,$$

$$\widetilde{\varphi}_R(\pi,\lambda) = 1, \quad \widetilde{\varphi}_R'(\pi,\lambda) = -h_\pi, \quad \text{hence} \quad \widetilde{B}_R\widetilde{\varphi}_R = 0.$$

We put

$$\Delta_R(\lambda) = \widetilde{\varphi}_R(x, \lambda)\varphi_R'(x, \lambda) - \widetilde{\varphi}_R'(x, \lambda)\varphi_R(x, \lambda),$$

and let

$$\lambda_0^{(R)} < \lambda_1^{(R)} < \cdots < \lambda_n^{(R)} < \cdots \tag{1.25}$$

be the zeros of $\Delta_R(\lambda)$. They are the simple eigenvalues of H_R, and $\varphi_R(x, \lambda_n)$ is an eigenvector associated with $\lambda_n^{(R)}$. Let $\alpha_n^{(R)} = \int_0^\pi \varphi_R(x, \lambda_n^{(R)})^2 dx$ be the normalizing constant. We also put[15]

$$\beta_n^{(R)} = \frac{\widetilde{\varphi}_R(x, \lambda_n^{(R)})}{\varphi_R(x, \lambda_n^{(R)})}. \tag{1.26}$$

Definition 1.2 We call $\{\lambda_n^{(R)}\}_{n=0}^\infty$, $\{\alpha_n^{(R)}\}_{n=0}^\infty$ the *spectral data* for H_R.

Lemma 1.5

$$\dot{\Delta}_R(\lambda_n^{(R)}) = \frac{d}{d\lambda}\Delta_R(\lambda)\Big|_{\lambda=\lambda_n^{(R)}} = -\alpha_n^{(R)}\beta_n^{(R)}, \text{ }[16] \tag{1.27}$$

$$\Delta_R(\lambda) = \pi\left(\lambda_0^{(R)} - \lambda\right)\prod_{n=1}^\infty \frac{\lambda_n^{(R)} - \lambda}{n^2}. \tag{1.28}$$

To prove (1.28), recall that the *order* of an entire function $f(z)$ is defined by

$$\kappa = \limsup_{r\to\infty} \frac{\log\log M(r)}{\log r}, \quad M(r) = \max_{|z|=r}|f(z)|.$$

If an entire function $f(z)$ is represented as

$$f(z) = Cz^m e^{g(z)} \prod_n \left(1 - \frac{z}{a_n}\right) e^{\frac{z}{a_n} + \frac{1}{2}(\frac{z}{a_n})^2 + \cdots + \frac{1}{h}(\frac{z}{a_n})^p}, \tag{1.29}$$

C being a constant, m a non-negative integer, and $g(z)$ a polynomial whose highest power is z^ℓ with $\ell \leq p$, $f(z)$ is said to be of finite order. The *genus* of $f(z)$ is

---

[15]Note that the right-hand side of (1.26) is a non-zero constant.
[16]Omitting the super (sub) script R, and integrating $\frac{d}{dx}\left(\widetilde{\varphi}(x, \lambda)\varphi'(x, \lambda_n) - \widetilde{\varphi}'(x, \lambda)\varphi(x, \lambda_n)\right) = (\lambda - \lambda_n)\widetilde{\varphi}(x, \lambda)\varphi(x, \lambda_n)$, we obtain $\Delta(\lambda) = (\lambda_n - \lambda)\int_0^\pi \widetilde{\varphi}(x, \lambda)\varphi(x, \lambda_n)dx$. Then, $\dot{\Delta}(\lambda_n) = -\int_0^\pi \widetilde{\varphi}(x, \lambda_n)\varphi(x, \lambda_n)dx$, which yields (1.27).

the smallest non-negative integer p such that $f(z)$ has the representation (1.29). Hadamard's factorization theorem [3, p. 207] implies that $p \leq \kappa \leq p+1$.

Omitting the subscript R, we have the integral equation

$$\varphi(x, \lambda) = \cos \sqrt{\lambda} x - h_0 \frac{\sin \sqrt{\lambda} x}{\sqrt{\lambda}} + \int_0^x \frac{\sin \sqrt{\lambda}(x-t)}{\sqrt{\lambda}} q(t) \varphi(t, \lambda) dt.$$

Then, we have as $\lambda \to \infty$

$$\varphi(x, \lambda) = \cos \sqrt{\lambda} x + q_1(x) \frac{\sin \sqrt{\lambda} x}{\sqrt{\lambda}} + \frac{1}{2\sqrt{\lambda}} \int_0^x q(t) \sin \sqrt{\lambda}(x - 2t) dt + O(e^{|\mathrm{Im}\sqrt{\lambda}|x} / \lambda),$$

where $q_1(x) = -h_0 + \frac{1}{2} \int_0^x q(t) dt$. This yields the asymptotic expansion

$$\Delta(\lambda) = -\sqrt{\lambda} \sin \sqrt{\lambda} \pi + \omega \cos \sqrt{\lambda} \pi + \frac{1}{2} \int_0^\pi q(t) \cos \sqrt{\lambda}(\pi - 2t) dt + O(e^{|\mathrm{Im}\sqrt{\lambda}|x} / \sqrt{\lambda}),$$

$$\omega = -h_0 + h_\pi + \frac{1}{2} \int_0^\pi q(t) dt. \tag{1.30}$$

Then $\Delta(\lambda)$ is an entire function of order $1/2$, and by Hadamard's factorization theorem, $\Delta(\lambda) = C \prod_{n=0}^\infty \left(1 - \frac{\lambda}{\lambda_n}\right)$. The constant C is computed by comparing with the case $V(x) = 0$: $\Delta_0(\lambda) = -\sqrt{\lambda} \sin \sqrt{\lambda} \pi = -\lambda \pi \prod_{n=1}^\infty \left(1 - \frac{\lambda}{n^2}\right)$.

For the Dirichlet case, i.e. $y(0) = 0$, $y'(\pi) + h_\pi y(\pi) = 0$, one can show that

$$\Delta_D(\lambda) = \prod_{n=0}^\infty \frac{\lambda_n^{(D)} - \lambda}{\left(n + \frac{1}{2}\right)^2}. \tag{1.31}$$

Lemma 1.6 *We have*[17]

$$M_R(\lambda) = -\frac{\Delta_D(\lambda)}{\Delta_R(\lambda)}. \tag{1.32}$$

By (1.28) and (1.31), $\{\lambda_n^{(R)}\}_{n=0}^\infty$ and $\{\lambda_n^{(D)}\}_{n=0}^\infty$ determine $M_R(\lambda)$. One can then compute $1/\alpha_n^{(R)}$ as the residue of $M_R(\lambda)$ at $\lambda = \lambda_n^{(R)}$. We thus have the following lemma.

Lemma 1.7 *Let* $\{\lambda_n^{(D)}\}_{n=0}^\infty$ *be the eigenvalues associated with the boundary condition* $y(0) = 0$, $\widetilde{B}_R y(\pi) = 0$. *Then,* $\{\lambda_n^{(R)}\}_{n=0}^\infty$ *and* $\{\lambda_n^{(D)}\}_{n=0}^\infty$ *determine* $\{\alpha_n^{(R)}\}_{n=0}^\infty$.

---

[17]Use $\psi_R(x, \lambda) = \widetilde{\varphi}_R(x, \lambda) / (\widetilde{\varphi}'_R(0, \lambda) + h_0 \widetilde{\varphi}_R(0, \lambda))$, $\Delta_R(\lambda) = \widetilde{\varphi}_R(x, \lambda) \varphi'_R(x, \lambda) - \widetilde{\varphi}'_R(x, \lambda) \varphi_R(x, \lambda)$, $\Delta_D(\lambda) = \widetilde{\varphi}_R(x, \lambda) \varphi'_D(x, \lambda) - \widetilde{\varphi}'_R(x, \lambda) \varphi_D(x, \lambda)$ and (1.6).

We derive relations between Weyl functions and spectral functions using the following theorem due to Nevanlinna [4, II, p. 7].

Theorem 1.9 *If $f(z)$ is analytic in $\mathbf{C}_+$ and satisfies $\operatorname{Im} f(z) \geq 0$ on $\mathbf{C}_+$, there exist unique constants $a \in \mathbf{R}$, $b \geq 0$ and a non-decreasing right-continuous function $\sigma(t)$ on $\mathbf{R}$ satisfying $\int_{-\infty}^{\infty} \frac{d\sigma(t)}{1+t^2} < \infty$, by which $f(z)$ is represented as*

$$f(z) = a + bz + \int_{-\infty}^{\infty} \frac{1+tz}{(t-z)(1+t^2)} d\sigma(t). \tag{1.33}$$

Conversely, $\sigma(t)$ is represented by $f(z)$. For $s < t$,

$$\frac{1}{2}\Big((\sigma(t)+\sigma(t-0)\Big) - \frac{1}{2}\Big((\sigma(s)-\sigma(s-0)\Big) = \lim_{\epsilon \downarrow 0} \frac{1}{\pi} \int_s^t \operatorname{Im} f(x+i\epsilon)dx. \tag{1.34}$$

Theorem 1.10

(1) $M(\lambda)$ is computed from $\rho(\lambda)$ as follows: For H_R,

$$- M(\lambda) = \int_{-\infty}^{\infty} \frac{d\rho(t)}{t-\lambda}, \tag{1.35}$$

For H_D, there exists a constant $a \in \mathbf{R}$ such that

$$M(\lambda) = a + \int_{-\infty}^{\infty} \frac{1+t\lambda}{(t-\lambda)(1+t^2)} d\rho(t). \tag{1.36}$$

(2) For both cases, $\rho(\lambda)$ is computed from $M(\lambda)$ as follows:

$$\frac{1}{2}\big(\rho(\lambda) + \rho(\lambda - 0)\big) = \mp \lim_{\epsilon \downarrow 0} \frac{1}{2\pi i} \int_{-\infty}^{\lambda} \big(M(t+i\epsilon) - M(t-i\epsilon)\big)dt,$$

where we take $-$ sign for H_R, and $+$ sign for H_D.[18]

We have thus seen that Weyl functions and spectral functions determine each other (up to a constant). In particular, they give spectral data $\{\lambda_n\}_{n=0}^{\infty}$ and $\{\alpha_n\}_{n=0}^{\infty}$.

---

[18]In fact, (1.35) is a direct consequence of (1.19). This agrees with (1.33) by taking $\sigma(t) = \rho(t)$, $b = 0$, and $a = \int_{-\infty}^{\infty} \frac{t}{1+t^2} d\rho(t)$. Note that $-\operatorname{Im} M_R(\lambda) \geq 0$. For the case H_D, $\operatorname{Im} M_D(\lambda) \geq 0$ by (1.7). By Theorem 1.9, there exists $\sigma(t)$ such that (1.36) holds. Observing $M_D'(\lambda)$ as $\operatorname{Im} \lambda \to \infty$, we see that $b = 0$. By (1.34), (1.19), and Lemma 1.2, we have $d\sigma(t) = \sum_{n=0}^{\infty} \frac{1}{\alpha_n} \delta(t - \lambda_n)dt = d\rho(t)$.

1.3 When Eigenvalues Determine the Potential?

The first result in the one-dimensional inverse boundary value problem is due to Ambarzumian [7].

Theorem 1.11 *Consider the eigenvalue problem* $-y'' + q(x)y = \lambda y$, $0 < x < \pi$, *with boundary condition* $y'(0) = y'(\pi) = 0$. *If* $\lambda_n = n^2$ *for all* $n \geq 0$, *then* $q(x) = 0$.

Let us prove this theorem. It is known that for $n \geq 1$, $\sqrt{\lambda_n} = n + \frac{\omega}{\pi n} + \frac{\kappa_n}{n}$, $\{\kappa_n\} \in \ell^2$. The assumption of the theorem implies $\int_0^\pi q(x)dx = 0$. Let $y_0(x)$ be the eigenfunction associated with the eigenvalue 0. Then, $y_0''(x) = q(x)y_0(x)$, $y_0'(0) = y_0'(1) = 0$. By the well-known Sturm's comparison theorem, $y_0(x)$ has no zeros on $[0, \pi]$. Since $\frac{y_0''(x)}{y_0(x)} = \left(\frac{y_0'(x)}{y_0(x)}\right)^2 + \left(\frac{y_0'(x)}{y_0(x)}\right)'$, we have

$$0 = \int_0^\pi q(x)dx = \int_0^\pi \frac{y_0''(x)}{y_0(x)}dx = \int_0^\pi \left\{\left(\frac{y_0'(x)}{y_0(x)}\right)^2 + \left(\frac{y_0'(x)}{y_0(x)}\right)'\right\}dx = \int_0^\pi \left(\frac{y_0'(x)}{y_0(x)}\right)^2 dx,$$

which implies $y_0'(x) = 0$, hence $y_0(x)$ is constant. By the equation, $q(x) = 0$.[19]

Therefore, for the case of the Neumann boundary condition, whether $q(x) = 0$ or not is determined only by the eigenvalues. However, this is exceptional, and one needs information of eigenfunctions or some other spectral data to determine the potential. Borg [26] proved that the knowledge of spectra from two different boundary conditions uniquely determines the potential.

Symmetry of the potential is another condition for the uniqueness. A potential $q(x)$ on $(0, 1)$ is said to be even if $q(x) = q(1 - x)$. Borg also showed that this evenness guarantees the uniqueness of the potential.

Theorem 1.12 *An even potential is uniquely determined from its eigenvalues.*[20]

---

[19]By the above proof one can see that if $\int_0^\pi q(x)dx = 0$ and 0 is the least eigenvalue of $H = -d^2/dx^2 + q$, then $q(x) = 0$. We give an alternative proof of this fact. By the assumption and the min-max principle, $\inf\{(u', u') + (qu, u) ; u \in H^1((0, \pi)), \|u\|_{L^2((0,1))} = 1\} = 0$. The assumption $\int_0^\pi q(x)dx = 0$ means that this infimum is attained by the constant function $u_0(x) = 1/\sqrt{\pi}$. Then, $u_0(x)$ is the eigenvector associated with the eigenvalue 0. The equation $-u_0'' + q(x)u = 0$ implies that $q(x) = 0$. This proof also works for n-dimensional compact manifolds M without or with boundary under the Neumann boundary condition. The assumption $\int_M q(x)\sqrt{g}dx = 0$ is proven by computing $\mathrm{tr}\,(e^{-tH} - e^{-tH_0})$ as $t \to 0$. See [131] and [58].

[20]See [62, p. 27] and [159, p. 117] for the proof.

1.4   Gel'fand–Levitan Equation

The function $\varphi_R(x, \lambda)$ is a solution to the initial value problem for the equation $-y'' + q(x)y - \lambda y = 0$ with initial data $y(0) = 1$, $y'(0) = -h_0$, which is transformed into an integral equation of Volterra type. Solving it by iteration, we obtain the following lemma, which is the key to the one-dimensional inverse spectral theory.

Lemma 1.8 *The function* $\varphi_R(x, \lambda)$ *has the following representation*

$$\varphi_R(x, \lambda) = \cos \sqrt{\lambda} x + \int_0^x G(x, t) \cos \sqrt{\lambda} t \, dt, \qquad (1.37)$$

where $G(x, t)$ *satisfies*

$$G(x, x) = -h_0 + \frac{1}{2} \int_0^x q(t) dt.$$

Therefore, the potential $q(x)$ is recovered from the kernel $G(x, t)$, and the issue of the inverse problem is reduced to this kernel. The idea consists in considering the integral operator

$$Tf(x) = f(x) + \int_0^x G(x, t) f(t) dt, \qquad (1.38)$$

called the *transformation operator*.

We have now prepared all necessary tools for the one-dimensional inverse problem. Let us summarize the algorithm for the recovery of the original boundary value problem from the spectral data $\{\lambda_n\}_{n=0}^\infty$, $\{\alpha_n\}_{n=0}^\infty$.

First note that they have the following asymptotic expansions for $n \geq 1$:

$$\sqrt{\lambda_n} = n + \frac{\omega}{n} + \frac{\kappa_n}{n}, \qquad \{\kappa_n\}_{n=1}^\infty \in \ell^2,$$

$$\alpha_n = \frac{\pi}{2} + \frac{\kappa_{n1}}{n}, \qquad \{\kappa_{n1}\}_{n=1}^\infty \in \ell^2,$$

where ω is the constant in (1.30). Define $F(x, t)$ by

$$F(x, t) = \sum_{n=0}^\infty \left(\frac{\cos \sqrt{\lambda_n} x \cos \sqrt{\lambda_n} t}{\alpha_n} - \frac{\cos nx \cos nt}{\alpha_n^{(0)}} \right),$$

$$\alpha_n^{(0)} = \begin{cases} \pi/2, & n > 0, \\ \pi, & n = 0. \end{cases}$$

Solve the *Gel'fand–Levitan equation*

$$G(x, t) + F(x, t) + \int_0^x G(x, s) \, F(s, t) \, ds = 0, \quad 0 < t < x.$$

Then, the potential $q(x)$ and the constants h_0, h_π are reconstructed as follows:

$$q(x) = 2 \frac{d}{dx} G(x, x), \quad h_0 = -G(0, 0), \quad h_\pi = \omega + h_0 - \frac{1}{2} \int_0^\pi q(t) dt.$$

We omit the details of the procedure for the reconstruction of the potential here, which can be seen in [62, 130, 131]. Instead, in Sect. 1.7 and Chap. 3, we will explain the mathematical background from the operator theoretical viewpoint for the scattering problem.

1.5 Spectral Mapping

There is another approach to the one-dimensional inverse problem developed by Trubowitz and others [51, 82, 83]. The main idea is to use the infinite dimensional complex analytic function theory, and regard the spectral data as the coordinates of some infinite dimensional analytic manifold. We follow [159] and consider the case of the Dirichlet problem. For the other boundary conditions, see [82] and [117].

1.5.1 Analytic Functions on a Banach Space

Let us briefly summarize the theory of complex analytic Banach manifolds, which is a direct extension of hypersurfaces in $\mathbf{C}^n$.

Let E, F be complex Banach spaces and U an open set in E. A function $f : U \to F$ is said to be *Frechét differentiable* at $q \in U$, if there exists a bounded linear operator $d_q f : E \to F$ having the following property. For any $\epsilon > 0$, there exists $\delta > 0$ such that

$$\| f(q + h) - f(q) - d_q f(h) \| \le \epsilon \|h\|, \quad \text{if} \quad \|h\| < \delta.$$

If f is Frechét differentiable at any point q in U, and $d_q f$ is continuous with respect to $q \in U$, i.e. for any $\epsilon > 0$, there exists $\delta > 0$ such that

$$\| d_{q'} f - d_q f \| < \epsilon, \quad \text{if} \quad \| q' - q \| < \delta,$$

where $\| d_{q'} f - d_q f \|$ is the operator norm while $\| q' - q \|$ is the norm in E, then f is said to be *analytic* on U. We emphasize here that E, F are complex Banach spaces.

When f is analytic, $d_q f$ is again Frechét differentiable with respect to q, and we obtain a symmetric bilinear operator

$$d_q^2 f : E \times E \to F, \quad d_q^2 f(u, v) = d_q^2 f(v, u).$$

This is continued, and as in the finite dimensional case, we obtain the Taylor series expansion

$$f(q + h) = f(q) + \sum_{n=1}^{\infty} \frac{1}{n!} d_q^n f(h, \cdots, h),$$

where $d_q^n f(\cdot, \cdots, \cdot)$ is a bounded symmetric multi-linear operator $\overbrace{E \times \cdots \times E}^{n} \to F$. Cauchy's formula holds: For $q, h \in E$ and $z \in \mathbf{C}$ such that $q + zh \in U$, we have

$$f(q + zh) = \frac{1}{2\pi i} \int_C \frac{f(q + \zeta h)}{\zeta - z} d\zeta,$$

where C is a sufficiently small contour in $\mathbf{C}$ enclosing z. Standard theorems in calculus are extended to the infinite dimensional case. In particular the following implicit function theorem holds.

Theorem 1.13 *Let E, F, G be complex Banach spaces and $U \subset E$, $V \subset F$ open sets. Let*

$$f : U \times V \to G$$

be analytic, and assume that at $(a, b) \in U \times V$, the partial derivative of f with respect to b

$$\partial_b f : F \to G$$

is a linear isomorphism. Then, there exists an open neighborhood $\mathcal{O}$ of a and a unique analytic function $g : \mathcal{O} \to V$ such that

$$f(x, g(x)) = f(a, b), \quad x \in \mathcal{O}, \quad \text{and} \quad g(a) = b.$$

A subset M in E is said to be an *analytic submanifold* of E, if for any $q \in M$, there exist

- an open neighborhood U of q,
- a complex Banach space F which is written as a direct sum $F = F_h \oplus F_v$, where F_h (horizontal space) and F_v (vertical space) are closed subspaces of F,
- an open set V in F,
- a diffeomorphism $\varphi : U \to V$ such that $\varphi(U \cap M) = V \cap F_h$.

We call $F_h \oplus F_v$ a splitting of F. The *tangent space* of M at q is defined by

$$T_q M = (d_q \varphi)^{-1}(F_h) = \{v \in E ; d_q \varphi(v) \in F_h\}.$$

Let E, F be Banach spaces and U an open set in E. For a continuously Frechét differentiable map $f : U \to F$, we put

$$M_c = \{q \in U ; f(q) = c\}.$$

A point $c \in F$ is said to be a *regular value* of f, if for any $q \in M_c$, there exists a splitting $E = E_h \oplus E_v$ such that

$$d_q f \big|_{E_v} : E_v \to F$$

is a linear isomorphism. We then have the following *regular value theorem*.

Theorem 1.14 *If $f : U \to F$ is an analytic map and $c \in F$ is a regular value, then M_c is an analytic submanifold, and $T_q M_c = \mathrm{Ker}\, d_q f$.*

For a real Hilbert space $\mathscr{H}_\mathbf{R}$, we define the real analyticity by introducing a complexification $\mathscr{H}_\mathbf{C}$ of $\mathscr{H}_\mathbf{R}$ and assuming that the mapping in question is extended to a complex analytic map on a complex neighborhood in $\mathscr{H}_\mathbf{C}$. Then, the above mentioned notions are extended to the real Hilbert space $\mathscr{H}_\mathbf{R}$.

We prepare one more notation. Let $L^2_\mathbf{R}(0, 1)$ be the set of real-valued L^2-functions on $(0, 1)$. The complexification of $L^2_\mathbf{R}(0, 1)$ is the usual $L^2(0, 1)$. Let $f : L^2_\mathbf{R}(0, 1) \to \mathbf{R}$ be a real analytic map. Then, for any $q \in L^2_\mathbf{R}(0, 1)$, $d_q(\cdot) : L^2_\mathbf{R}(0, 1) \to \mathbf{R}$ is a continuous linear functional. By Riesz' theorem, there exists a $g(x) \in L^2_\mathbf{R}(0, 1)$ such that

$$d_q f(v) = \int_0^1 v(x) g(x) dx, \quad \forall v \in L^2_\mathbf{R}(0, 1).$$

We denote this $g(x)$ as $\dfrac{\partial f}{\partial q}(x)$. Therefore,

$$d_q f(v) = \int_0^1 v(x) \frac{\partial f}{\partial q}(x) dx, \quad \forall v \in L^2_\mathbf{R}(0, 1).$$

Let us also note that we have

$$d_q f(v) = \frac{d}{d\epsilon} f(q + \epsilon v)\big|_{\epsilon=0},$$

the right-hand side of which is called *Gâteaux derivative*.

1.5.2 Analytic Map Associated with the Eigenvalue Problem

Let us apply the above abstract theory to the boundary value problem. We explain only the introductory part. Let $y(x, \lambda, q)$ be the solution of

$$-y'' + q(x)y = \lambda y, \quad 0 < x < 1, \tag{1.39}$$

satisfying the initial condition

$$y(0, \lambda, q) = 0, \quad y'(0, \lambda, q) = 1.$$

First let us note that $y(x, \lambda, q)$ is analytic with respect to $q(x) \in L^2(0, 1)$. In fact, $y(x, \lambda)$ is constructed in the form

$$y(x, \lambda, q) = s_\lambda(x) + \sum_{n=1}^{\infty} S_n(x, \lambda, q),$$

where $s_\lambda(x) = \dfrac{\sin \sqrt{\lambda} x}{\sqrt{\lambda}}$, and $S_n(x, \lambda, q)$ is defined by

$$-S_n'' = \lambda S_n - q(x)S_{n-1}, \quad S_n(0) = 0, \quad S_n'(0) = 0.$$

Then, we have

$$S_n(x, \lambda, q) = \int_{0 \leq t_1 \leq \cdots \leq t_{n+1} = x} t_i \prod_{i=1}^{n} (t_{i+1} - t_i)\big(q(t_i) - \lambda\big) dt_1 \cdots dt_n,$$

which shows that $y(x, \lambda, x)$ is expanded as a power series with respect to q, hence it is analytic.

We consider the Dirichlet eigenvalue problem for the Eq. (1.39). Then, λ is an eigenvalue if and only if $y(1, \lambda, q) = 0$. Let

$$\lambda_1(q) < \lambda_2(q) < \cdots$$

be the eigenvalues. One can show that[21]

$$\dot{y}(1, \lambda, q)y'(1, \lambda, q) = \int_0^1 y(x, \lambda, q)^2 dx.$$

---

[21]Differentiating (1.39) with respect to λ, we have $-\dot{y}'' + q\dot{y} = y + \lambda \dot{y}$. We then have $(\dot{y}y' - \dot{y}y')' = y^2$, integrating which we obtain this equality.

Therefore, $\dot{y}(1, \lambda, q) \neq 0$, which implies that $\lambda_n(q)$ is a real-analytic function of $q(x) \in L^2_{\mathbf{R}}(0, 1)$. Let

$$g_n(x, q) = \frac{y(x, \lambda_n(q), q)}{\|y(\cdot, \lambda_n(q), q)\|},$$

where $\| \cdot \|$ is the norm of $L^2(0, 1)$. Then, the Frechét derivative of $\lambda_n(q)$ is written as follows.[22]

$$\frac{\partial \lambda_n}{\partial q}(x) = g_n^2(x, q). \tag{1.40}$$

For $q(x) \in L^2_{\mathbf{R}}(0, 1)$, the eigenvalue $\lambda_n(q)$ has the following asymptotic expansion

$$\lambda_n(q) = (n\pi)^2 + \int_0^1 q(x)dx + \mu_n(q), \tag{1.41}$$

where $\mu_n(q)$ satisfies

$$\sum_{n=1}^{\infty} (\mu_n(q))^2 < \infty.$$

We put

$$\mu(q) = (\mu_0(q), \mu_1(q), \mu_2(q), \cdots), \quad \mu_0(q) = \int_0^1 q(x)dx.$$

Then, μ is a map:

$$\mu : L^2_{\mathbf{R}}(0, 1) \rightarrow \mathbf{R} \times \ell^2_{\mathbf{R}} =: S,$$

where $\ell^2_{\mathbf{R}}$ is the Hilbert space of square summable real sequences. By (1.40) and (1.41), we have the following formula:

$$d_q\mu(v) = \left(\int_0^1 v(t)dt, \int_0^1 v(t)(g_1^2(t) - 1)dt, \int_0^1 v(t)(g_2^2(t) - 1)dt, \cdots \right). \tag{1.42}$$

---

[22]Taking the Frechét derivative of (1.39), we have $-(d_q g_n(v))'' + v g_n + q d_q g_n(v) = (d_q \lambda_n)(v)g_n + \lambda_n d_q g_n(v)$. Multiplying by g_n and integrating by parts, we have $(d_q g_n(v), (-\frac{d^2}{dx^2} + q)g_n) + (g_n^2, v) = (d_q \lambda_n)(v) + \lambda_n(d_q g_n(v), g_n)$, which implies the formula.

We put

$$\kappa_n(q) = \log\left((-1)^n y'(1, \lambda_n(q), q)\right),$$

$$\kappa(q) = (\kappa_1(q), \kappa_2(q), \kappa_3(q), \cdots).$$

Define the real Hilbert space ℓ_1^2 by

$$\ell_1^2 \ni \alpha = (\alpha_1, \alpha_2, \cdots) \iff \sum_{n=1}^{\infty} (n\alpha_n)^2 < \infty.$$

Let $y_i(x, \lambda, q)$, $i = 1, 2$, be solutions to (1.39) satisfying the initial conditions

$$y_1(0, \lambda, q) = 1, \quad y_1'(0, \lambda, q) = 0,$$

$$y_2(0, \lambda, q) = 0, \quad y_1'(0, \lambda, q) = 1,$$

and put $a_n(x, q) = y_1(x, \lambda_n(q), q)y_2(x, \lambda_n(q), q)$. The Frechét derivative of κ_n is computed as follows:

$$\frac{\partial \kappa_n}{\partial q}(x) = a_n(x, q) - g_n^2(x, q) \int_0^1 a_n(t, q)dt. \qquad (1.43)$$

By (1.40) and (1.43), one has

Lemma 1.9

$$\frac{\partial \mu_n}{\partial q}(x) = -\cos 2\pi nx + O\left(\frac{1}{n}\right),$$

$$\frac{\partial \kappa_n}{\partial q}(x) = \frac{1}{2\pi n}\sin 2\pi nx + O\left(\frac{1}{n^2}\right).$$

A set of vectors $\{v_n\}_{n=1}^{\infty}$ in a Hilbert space $\mathcal{H}$ is said to be linearly independent if $v_n \notin \mathcal{H}_n$ for all $n \geq 1$, where $\mathcal{H}_n$ is the closure of the linear hull of $\{v_m ; \ m \neq n\}$

Lemma 1.10 *Let* $\{e_n\}_{n=1}^{\infty}$ *be a complete orthonormal system of a Hilbert space* $\mathcal{H}$. *Suppose* $\{d_n\}_{n=1}^{\infty} \subset \mathcal{H}$ *spans* $\mathcal{H}$ *or is linearly independent. If, in addition,* $\sum_{n=1}^{\infty} \|d_n - e_n\|^2 < \infty$, *then* $\{d_n\}_{n=1}^{\infty}$ *is also a basis of* $\mathcal{H}$. *Moreover, the map*

$$\mathcal{H} \ni x \to ((x, d_1), (x, d_2), \cdots) \in \ell^2$$

is a linear isomorphism.

By Lemmas 1.9 and 1.10, $\kappa \times \mu$ is a local diffeomorphism. To prove that it is a global isomorphism, one needs a reconstruction procedure of the potential, which is the content of inverse problem. One thus obtains the following theorem.

Theorem 1.15 *The map* $\kappa \times \mu : L_{\mathbf{R}}^2(0, 1) \to \ell_1^2 \times S$ *is a real analytic isomorphism.*

This theorem gives a characterization of a pair of sequences $(\kappa_1, \kappa_2, \cdots)$ and $(\mu_0, \mu_1, \dots)$ to become the spectral data of a potential in $L_{\mathbf{R}}^2(0, 1)$.

For $q \in L_{\mathbf{R}}^2(0, 1)$, the isospectral set of potentials is defined by

$$M(q) = \left\{ V \in L_{\mathbf{R}}^2(0, 1) \, ; \, \lambda_n(q) = \lambda_n(V), \ \forall n \geq 1 \right\}.$$

Theorem 1.16 $M(q)$ *is a real analytic submanifold of* $L_{\mathbf{R}}^2(0, 1)$, *and* κ *is a global coordinate on* $M(q)$.

This theorem enables us to deform the potential by varying the data κ keeping the eigenvalues invariant.

1.5.3 Liouville Transformation

The Liouville transformation $f \to \rho f$ maps the Sturm–Liouville equation of *impedance form*

$$- \rho^{-2} (\rho^2 f')' + u f = \lambda f, \tag{1.44}$$

to the Schrödinger equation

$$- y'' + p y = \lambda y. \tag{1.45}$$

Transforming back, in principle, one can solve the inverse problem for the Sturm–Liouville equation of impedance form (1.44) utilizing the results for the Schrödinger equation (1.45). However, it is not obvious to find a proper form of the associated transformation between the function spaces for impedance functions ρ in (1.44) and potentials p in (1.45). In [89] and [90] (see also [101, 102]), it is shown that there exists an analytic isomorphism between a space of impedance functions ρ and that for the potentials p, which also preserves the boundary conditions and the spectral data, i.e. eigenvalues and norming constants. It then gives an isomorphism between the solutions of inverse problems for these equations.

An example of the Sturm–Liouville equation of impedance form is given by the surface of revolution, which is obtained by rotating a curve $z = f(x)$ around x-axis. Letting M be the resulting surface in $\mathbf{R}^3$, the induced metric on M is

$$ds^2 = (1 + f'(x)^2)(dx)^2 + f(x)^2 g_{S^1},$$

where g_{S^1} is the standard metric on S^1. Representing S^1 as $\{(\cos\theta, \sin\theta) ; 0 \leq \theta \leq 2\pi\}$, we have $g_{S^1} = (d\theta)^2$. Making the change of variable $x \to t$ by $dt/dx = \sqrt{1 + f'(x)^2}$, we have $ds^2 = (dt)^2 + r(t)^2(d\theta)^2$, where $r(t) = f(x(t))$. Then, the Laplacian on M is $\Delta_M = r^{-1}\partial_t(r\partial_t) + r^{-2}\partial_\theta^2$, which is self-adjoint under a suitable boundary condition. By the Fourier series expansion, we are reduced to the operator $H_n = -r^{-1}\partial_t(r\partial_t) + r^{-2}n^2$. Fixing $n \in \mathbf{Z}$ arbitrarily, we have arrived at a one-dimensional inverse problem : To determine the manifold M, i.e. $f(x)$, from the spectral data of H_n.

Brüning–Heintze [29] applied the one-dimensional Gel'fand–Levitan theory to solve this problem. Zelditch [200] proved that the isospectral surfaces of revolution of simple length spectrum, with some additional conditions, are isometric. In [89] and [90], an analytic isomorphism from the space of spectral data onto the space of functions describing the radius of rotation was constructed.

The Minkowski problem in classical differential geometry asks the existence of a surface in $\mathbf{R}^3$ with a prescribed Gaussian curvature. It was solved by Pogorelov [158] and Cheng–Yau [39]. In the case of the surface of revolution, the Gaussian curvature G is written as

$$G(x) = -(2q(x))' - (2q(x))^2, \quad q(x) = \frac{r'(x)}{2r(x)}.$$

Therefore, one can solve the Minkowski problem in the case of the surface of revolution by constructing an analytic isomorphism between the space of Gaussian curvature $G(x)$ and the space of profiles $r(x)$. See [90].

1.6 Isospectral Deformation

It is well-known that for two bounded operators A and B

$$\sigma(AB) \setminus \{0\} = \sigma(BA) \setminus \{0\}, \tag{1.46}$$

which is proven by using the relation

$$\lambda(AB + \lambda)^{-1} + A(BA + \lambda)^{-1}B = 1.$$

For unbounded operators, (1.46) is also true if A is a densely defined closed operator and $B = A^*$. This is a hidden mathematical basis of the one-dimensional inverse problem. In fact, by a straightforward computation, Crum [46] used this method to deform and remove eigenvalues of Sturm–Liouville operators. This commutation method has a long history, precursors of which are seen in the works of Jacobi [97] and Darboux [52] Let us formally elucidate it here.

Let $V(x)$ be a real-valued potential and φ_n a Dirichlet eigenfunction of $-d^2/dx^2 + V(x)$ with eigenvalue λ_n. Ignoring the domain question of all relevant operators, we put formally

$$A = \varphi_n \frac{d}{dx}\left(\frac{1}{\varphi_n}\cdot\right), \quad A^* = -\frac{1}{\varphi_n}\frac{d}{dx}\left(\frac{1}{\varphi_n}\cdot\right).$$

Using $-\varphi_n'' + V\varphi_n = \lambda_n$, we then have

$$A^*A = -\frac{d^2}{dx^2} + V(x) - \lambda_n,$$

$$AA^* = -\frac{d^2}{dx^2} + V(x) - 2\frac{d^2}{dx^2}\log\varphi_n(x) - \lambda_n.$$

Therefore, we will have $\sigma(A^*A) \setminus \{0\} = \sigma(AA^*) \setminus \{0\}$. However, the potential term of AA^* is not in L^2 in any neighborhoods of the zeros of $\varphi_n(x)$, which causes a difficulty. The remedy comes from a double commutation. Putting

$$u = \frac{1}{\varphi_n}\left(a + b\int_0^x \varphi_n(t)^2 dt\right),$$

$$B = u\frac{d}{dx}\left(\frac{1}{u}\cdot\right), \quad B^* = -\frac{1}{u}\frac{d}{dx}\left(u\cdot\right),$$

we have

$$BB^* = -\frac{d^2}{dx^2} + V(x) - 2\frac{d^2}{dx^2}\log\left(u\varphi_n\right) - \lambda_n,$$

moreover

$$BB^*\frac{1}{u} = 0.$$

This shows that $V(x)$ and $V(x) - 2\frac{d^2}{dx^2}\log\left(u\varphi_n\right)$ have the same Dirichlet eigenvalues $\{\lambda_n\}_{n=1}^\infty$.

This commutation method does not work in multi-dimension. In the next chapter, we show that there is no analogue of isospectral deformation of $-\Delta + V(x)$ as in Theorem 1.16.

1.7   Scattering on the Half-Line

We consider the scattering problem for Schrödinger operators in $\mathbf{R}^3$. We postpone the detailed explanation until Chap. 3, and summarize here its basic facts. Let

$$H_0 = -\Delta, \quad H = H_0 + V(x),$$

where $V(x)$ is real-valued and $V(x) \to 0$ as $|x| \to \infty$. As in the case of one-dimension, $\mathscr{H}_{ac}(H)$ is expanded by the generalized plane wave $\varphi(x, \xi)$ which satisfies

$$(-\Delta + V(x))\varphi(x, \xi) = |\xi|^2 \varphi(x, \xi),$$

and behaves like

$$\varphi(x, \xi) \sim e^{ix \cdot \xi} + \frac{e^{i|\xi|r}}{r} a(|\xi|, \widehat{x}, \widehat{\xi}), \quad r = |x| \to \infty,$$

where $\widehat{x} = x/r$. The first term in the right-hand side represents the incoming plane wave, and the second term the scattered spherical wave. Letting $k = |\xi|$, $\theta = \widehat{x}$, $\omega = \widehat{\xi}$, the scattering matrix $S(k)$ (or *Heisenberg's S-matrix*) is written as

$$(S(k)\psi)(\theta) = \psi(\theta) - \int_{S^2} a(k, \theta, \omega)\psi(\omega)d\omega,$$

$$a(k, \theta, \omega) = -\frac{1}{4\pi} \int_{\mathbf{R}^3} e^{-ik\theta \cdot x} V(x)\varphi(x, k\omega)dx.$$

We call $a(k, \theta, \omega)$ the *scattering amplitude*, and $|a(k, \theta, \omega)|^2$ the *differential cross section*. Given a beam of particles coming from the direction ω, it gives the ratio of number of particles scattered to the direction θ. When $V(x)$ decays sufficiently rapidly at infinity, the scattering amplitude has an analytic continuation with respect to k onto the upper half-plane $\{\text{Im}\,k > 0\}$ with poles on the imaginary axis $k_1 = i\gamma_1, k_2 = i\gamma_2, \cdots$. Moreover, $k_j^2 = -\gamma_j^2$ are the negative eigenvalues of $-\Delta + V(x)$.

Assume that $V(x)$ is spherically symmetric : $V(x) = q(|x|)$. Letting

$$\omega = (0, 0, 1), \quad x \cdot \omega = r\cos\theta, \quad r = |x|,$$

the above generalized plane wave is expanded as follows:

$$\varphi(x, k\omega) = \frac{4\pi}{kr} \sum_{\ell=0}^{\infty} (2\ell + 1)i^\ell P_\ell(\cos\theta)\psi_\ell(k, r),$$

where ψ_ℓ satisfies

$$\begin{cases} \left(-\dfrac{d^2}{dr^2} + \dfrac{\ell(\ell+1)}{r^2} + q(r)\right)\psi_\ell = k^2 \psi_\ell, \\ \psi_\ell(k, 0) = 0. \end{cases}$$

As $r \to \infty$, it has the following asymptotic expansion

$$\psi_\ell(k, r) \sim e^{i\delta_\ell} \sin\left(kr - \frac{\ell\pi}{2} + \delta_\ell\right) = \frac{1}{2i}\left(e^{i(kr - \frac{\ell\pi}{2})}e^{2i\delta_\ell} - e^{-i(kr - \frac{\ell\pi}{2})}\right),$$

where $\delta_\ell = \delta_\ell(k)$ is called the *phase shift*, and

$$S_\ell(k) = e^{2i\delta_\ell(k)}$$

is the *S-matrix*. The scattering amplitude in $\mathbf{R}^3$ is rewritten as follows:

$$a(k, \theta, \omega) = \frac{1}{k} \sum_{\ell=0}^{\infty} (2\ell + 1)e^{i\delta_\ell(k)}\left(\sin \delta_\ell(k)\right)P_\ell(\theta \cdot \omega).$$

As is seen above, the scattering amplitude $a(k, \theta, \omega)$ has a meromorphic extension onto the upper half-plane with poles on the imaginary axis : $k_1 = i\gamma_1, k_2 = i\gamma_2, \cdots$ with $\gamma_j > 0$. Moreover $k_j^2 = -\gamma_j^2$ is a negative eigenvalue of H. Therefore, each phase shift $\delta_\ell(k)$ has the same property.

For the sake of simplicity, we restrict ourselves to the case $\ell = 0$, and consider the equation

$$Hy = -y'' + q(x)y = k^2 y, \quad 0 < x < \infty, \tag{1.47}$$

$$y(0) = 0. \tag{1.48}$$

The one-dimensional inverse scattering problem was solved by Gel'fand–Levitan–Marchenko [64, 137]. The original work of Gel'fand–Levitan was formulated by the spectral function. The complete correspondence with the S-matrix was studied by Marchenko (see p. 218, and p. 234 Theorem 3.3.3 of [138]) by giving not only the reconstruction procedure of the potential but also necessary and sufficient conditions for a function $S(k)$ to be the S-matrix for a Schrödinger operator $-\frac{d^2}{dx^2} + q(x)$.

Theorem 1.17 *In order that a function $S(k)$ defined for $k \in \mathbf{R}$ is the S-matrix of a Schrödinger operator $-\frac{d^2}{dx^2} + q(x)$ on $(0, \infty)$ with Dirichlet condition at $x = 0$ and the potential $q(x)$ satisfying $\int_0^\infty x|q(x)|dx < \infty$, it is necessary and sufficient that the following three conditions are satisfied.*

(1) $S(k)$ is continuous, $|S(k)| = 1$, $\overline{S(k)} = S(-k)$ for $k \in \mathbf{R}$.
(2) $S(k) - 1$ is written as

$$S(k) - 1 = \int_{-\infty}^{\infty} e^{-ikt}\left(F_1(t) + F_2(t)\right)dt,$$

where

$$F_1(t) \in L^1(\mathbf{R}), \quad F_2(t) \in L^2(\mathbf{R}) \cap L^\infty(\mathbf{R}), \quad \int_0^\infty t \left| F_1'(t) + F_2'(t) \right| dt < \infty.$$

(3) There exists an integer $m \geq 0$ such that

$$\arg S(+0) - \arg S(+\infty) + \frac{\pi}{2} \left(S(0) - 1 \right) = 2\pi m.$$

If $m = 0$, *the potential* $q(x)$ *is uniquely determined. If* $m > 0$, *there exists an m-parameter family of potentials having the same* $S(k)$ *as its S-matrix.*

1.7.1 Generalized Sine Transform

Let us study the mathematical background of the one-dimensional inverse scattering theory. Let $\varphi(x, k)$ be the solution of the equation

$$\begin{cases} -\varphi''(x, k) + q(x)\varphi(x, k) = k^2\varphi(x, k), & x > 0, \\ \varphi(0, k) = 0, \quad \varphi'(0, k) = 1. \end{cases} \tag{1.49}$$

It has the following behavior as $|x| \to \infty$:

$$\varphi(x, k) = \frac{|F(k)|}{k} \sin(kx + \delta(k)) + o(1),$$

where $F(k) = |F(k)|e^{-i\delta(k)}$. We are going to relate $F(k)$ to the spectral data of $H = -\frac{d^2}{dx^2} + q(x)$. For $|k| \to \infty$, $\varphi(x, k)$ behaves like

$$\varphi(x, k) = \frac{\sin kx}{k} (1 + o(1)).$$

We define the spectral measure by

$$\rho(\lambda) = \int_{-\infty}^\lambda d\rho(\lambda),$$

$$\frac{d\rho(\lambda)}{d\lambda} = \begin{cases} \dfrac{1}{\pi} \dfrac{\sqrt{\lambda}}{|F(\sqrt{\lambda})|^2}, & \lambda \geq 0, \\[2ex] \displaystyle\sum_{j=1}^m C_j \delta(\lambda - \lambda_j), & \lambda < 0, \end{cases}$$

where $\lambda_j = k_j^2 = -\gamma_j^2$ is the negative eigenvalue of H, and

$$C_j^{-1} = \int_0^\infty \varphi_j(x)^2 dx, \quad \varphi_j(x) = \varphi(x, k_j).$$

We define

$$(\mathscr{F}u)(\lambda) = \begin{cases} \displaystyle\int_0^\infty u(x)\varphi(x, \sqrt{\lambda})dx, & \lambda > 0, \\ \displaystyle\int_0^\infty u(x)\varphi_j(x)dx, & \lambda = \lambda_j. \end{cases} \tag{1.50}$$

Then, any $u \in L^2((0, \infty))$ is expanded as

$$u(x) = \mathscr{F}^*\mathscr{F}u$$

$$= \int_0^\infty \varphi(x, \sqrt{\lambda})(\mathscr{F}u)(\lambda)d\rho(\lambda) + \sum_j C_j(u, \varphi_j)\varphi_j(x). \tag{1.51}$$

1.7.2 The Core of Gel'fand–Levitan Theory

Recall that an entire function $f(z)$ is of *exponential type* σ if for any $\epsilon > 0$, there exists a constant $C_\epsilon > 0$ such that

$$|f(z)| \le C_\epsilon e^{(\sigma+\epsilon)|z|}, \quad \forall z \in \mathbf{C}.$$

Theorem 1.18 (Paley–Wiener) $f(x) \in L^2(\mathbf{R})$ *is extended to an entire function of exponential type* $\sigma > 0$ *if and only if there exists* $h \in L^2(-\sigma, \sigma)$ *such that*

$$f(x) = \int_{-\sigma}^\sigma h(\xi)e^{ix\xi}d\xi.$$

Let $\varphi(x, k)$ be as in (1.49). Then, $\varphi(x, k)$ is even and entire in $k \in \mathbf{C}$. By the Paley–Wiener theorem, it is rewritten as

$$\varphi(x, k) = \frac{\sin kx}{k} + \int_0^x K(x, y)\frac{\sin ky}{k}dy.$$

The main steps of the Gel'fand–Levitan–Marchenko theory are the following. This K is shown to satisfy

$$(\partial_y^2 - \partial_x^2 + V(x))K(x, y) = 0.$$

The Gel'fand–Levitan equation holds:

$$K(x, y) + G(x, y) + \int_0^x K(x, t)G(t, y)dt = 0, \quad x > y > 0, \tag{1.52}$$

where $G(x, y)$ is constructed from the S-matrix and the bound states to be defined later. Moreover, K satisfies

$$2\frac{d}{dx}(K(x, x)) = q(x). \tag{1.53}$$

Therefore, the inverse scattering consists of three procedures: Given a phase shift $\delta(k)$, we are going to do the followings.

(1) The construction of $G(x, y)$.
(2) Solving the Gel'fand–Levitan equation (1.52).
(3) Computation of $q(x)$ by (1.53).

1.7.3 Jost Solution and Spectral Data

We look at the construction of $G(x, y)$. Recall that Jost solution $f(x, k)$ defined by (1.12). Another definition of the Jost solution is a solution to (1.47) satisfying the boundary condition at infinity:

$$f(x, k) = e^{ikx}(1 + o(1)), \quad x \to \infty.$$

Let $C(x, k)$, $S(x, k)$ be solutions to (1.47) with boundary condition

$$C(0, k) = 1, \quad C'(0, k) = 0,$$

$$S(0, k) = 0, \quad S'(0, k) = 1.$$

We let

$$\Phi_0(x, k) = \frac{f(x, k)}{f(0, k)} = C(x, k) + M_0(k)S(x, k). \tag{1.54}$$

Then, $\Phi_0(x, k)$ satisfies $-y'' + q(x)y = k^2 y$ and

$$\Phi_0(0, k) = 1, \quad \Phi_0'(0, k) = M_0(k).$$

If $q \in L^2((0, 1))$, we have

$$M_0(k) = \frac{f'(0, k)}{f(0, k)}.$$

Therefore, $M_0(k)$ is the Weyl function (see the end of Sect. 1.1.2).

The Jost solution $f(x, k)$ is related with $\varphi(x, k)$ as follows:

$$\varphi(x, k) = \frac{1}{2ik} (F(-k)f(x, k) - F(k)f(x, -k)),$$

$$F(k) = W(f(x, k), \varphi(x, k)). \tag{1.55}$$

For $k \in \mathbf{R} \setminus \{0\}$, we have the following expansion as $x \to \infty$

$$\psi(x, k) := \frac{k}{F(k)} \varphi(x, k) \sim e^{i\delta(k)} \sin(kx + \delta(k)) = \frac{1}{2i} \left(e^{2i\delta(k)} e^{ikx} - e^{-ikx} \right). \tag{1.56}$$

The formula (1.56) is interpreted as follows : Given the incident wave e^{-ikx}, one observes the scattered wave e^{ikx}, the coefficient of which is the phase shift.

Therefore, this $\psi(x, k)$ is the solution to $-y'' + q(x)y = k^2 y$ appearing in the expansion

$$\Psi(\mathbf{x}, \mathbf{k}) = \frac{4\pi}{kr} \sum_{\ell=0}^{\infty} (2\ell + 1)i^\ell P_\ell(\cos\theta)\psi_\ell(k, r), \quad |\mathbf{x}| = r$$

of the solution to the three-dimensional Schrödinger equation.

There are two routes for the passage from $\delta(k)$ to $F(k)$, both of which rely on complex analysis. Recall that the Hilbert transform $\mathscr{H}$ is defined by

$$\mathscr{H} f = \text{p.v.} \frac{1}{\pi} \int_{-\infty}^{\infty} \frac{f(y)}{x - y} dy.$$

Passing to the Fourier transform, it is rewritten as

$$(\widehat{\mathscr{H} f})(\xi) = i \, \text{sgn}(\xi) \widehat{f}(\xi).$$

Therefore, $\mathscr{H}$ is unitary on $L^2(\mathbf{R})$ and

$$\mathscr{H}^2 = -1.$$

The *Hardy class* H_p, $p > 0$, is the set of analytic functions $F(z)$ on $\mathbf{C}_+$ such that

$$\sup_{y>0} \int_{-\infty}^{\infty} |F(x+iy)|^p dy < \infty.$$

If $F(z) \in H_p$ with $p \geq 1$, $F(x+i\epsilon) \to f(x)$, a.e. and $f(x) \in L^p(\mathbf{R})$. Moreover

$$F(z) = \frac{1}{2\pi i} \int_{-\infty}^{\infty} \frac{f(t)}{t-z} dt, \quad z \in \mathbf{C}_+.$$

Theorem 1.19 *If $F(z) \in H_p$ with $p \geq 1$, and $F(x+i\epsilon) \to f(x)$, we have*

$$\operatorname{Im} f = \mathscr{H} \operatorname{Re} f, \quad \operatorname{Re} f = -\mathscr{H} \operatorname{Im} f.$$

For the proof, see [116, p. 129].

We construct $F(k)$ from $\delta(k)$ and $\{\gamma_j\}$, which should satisfy

$$F(k) = |F(k)|e^{-i\delta(k)}, \quad \operatorname{Im} k = 0. \tag{1.57}$$

$$S(k) = \frac{F(-k)}{F(k)} = e^{2i\delta(k)}.$$

Assuming that we have constructed $F(k)$, we remove the poles $\{i\gamma_j\}$ from $F(k)$, where $-\gamma_j^2 = \lambda_j$ is a negative eigenvalue of $d^2/dx^2 + q(x)$. Let for $k \in \mathbf{C}_+$

$$G(k) = \prod_{j=1}^{m} \left(\frac{k+i\gamma_j}{k-i\gamma_j}\right) F(k),$$

$$\widetilde{\delta}(k) = 2 \sum_{j=1}^{m} \arctan \frac{\gamma_j}{k} + \delta(k).$$

Then, we have for $k \in \mathbf{R}$

$$\operatorname{Re} \log G(k) = \text{p.v.} \frac{1}{\pi} \int_{-\infty}^{\infty} \frac{\widetilde{\delta}(k')}{k-k'} dk'.$$

Then, $F(k)$ has the form

$$F(k) = \prod_{j} \left(1 + \frac{\gamma_j^2}{k^2}\right) \exp\left(\frac{1}{\pi} \int_{-\infty}^{\infty} \frac{\delta(k')}{k-k'} dk'\right), \quad \operatorname{Im} k > 0,$$

Letting $\operatorname{Im} k \to 0$, we have

$$|F(k)| = \prod_j \left(1 + \frac{\gamma_j^2}{k^2}\right) \exp\left(\frac{1}{\pi} \text{p.v.} \int_{-\infty}^{\infty} \frac{\delta(k')}{k - k'} dk'\right), \quad \operatorname{Im} k = 0, \tag{1.58}$$

The formulas (1.58) and (1.57) are the representation of $F(k)$ in terms of $\delta(k)$. Another route uses Wiener-Levi's theorem.

Theorem 1.20 *Let $A(z)$ be analytic in a domain $D \subset \mathbf{C}$, and assume that*

$$F(k) = C_F + \int_{-\infty}^{\infty} e^{ikt} f(t) dt,$$

where C_F is a constant and $f(t) \in L^1(\mathbf{R})$, satisfies $F(k) \in D$ for all $k \in \mathbf{R}$. Then, there exists $g(t) \in L^1(\mathbf{R})$ such that

$$A(F(k)) = C_A + \int_{-\infty}^{\infty} e^{ikt} g(t) dt,$$

where $C_A = A(C_F)$.

For the proof, see, e.g., [67], p.29, 9D. Theorem.[23]
We put

$$G(k) = \prod_{j=1}^{m} \left(\frac{k + i\gamma_j}{k - i\gamma_j}\right) F(k).$$

Then, by Wiener-Levi's theorem, $\log G(k)$ is written as

$$\log G(k) = \int_0^{\infty} e^{ikt} g(t) dt, \quad g(t) \in L^1(0, \infty),$$

since $F(k)$ is written as

$$F(k) = 1 + \int_0^{\infty} e^{ikt} a(t) dt.$$

---

[23] As a matter of fact, in addition to this theorem, we need the following lemma.

Lemma 1.11 *Let D be a domain in $\mathbf{C}$ such that $0 \in D$, and suppose $\phi(z)$ is analytic in D and satisfies $\phi(0) = 0$. Then, for any $f \in L^1(\mathbf{R})$, there exist $R > 0$ and $g \in L^1(\mathbf{R})$ such that $\phi(\widehat{f}(k)) = \widehat{g}(k)$ for $|k| > R$.*

We then have

$$F(k) = \prod_{j=1}^{m} \left(\frac{k + i\gamma_j}{k - i\gamma_j}\right) \exp\left(\int_0^\infty e^{ikt} g(t)dt\right), \tag{1.59}$$

$$\delta(k) = -\text{Im} \log F(k) = i \sum_{j=1}^{m} \log\left(\frac{k + i\gamma_j}{k - i\gamma_j}\right) - \int_0^\infty e^{ikt} g(t)dt. \tag{1.60}$$

The formulas (1.59) and (1.60) are another representation of $F(k)$ by $\delta(k)$.

Finally, $G(x, y)$ is represented as follows.

$$G(x, y) = \int_{-\infty}^{\infty} \frac{\sin\sqrt{\lambda}x \sin\sqrt{\lambda}y}{\sqrt{\lambda}} \, d\sigma(\lambda), \tag{1.61}$$

where the spectral function $\sigma(\lambda)$ is constructed by the formula

$$d\sigma(\lambda) = \begin{cases} \dfrac{1}{\pi}\left(\dfrac{1}{|F(\sqrt{\lambda})|^2} - 1\right)\sqrt{\lambda}d\lambda, & \lambda \geq 0, \\[2mm] \displaystyle\sum_{j=1}^{n} C_j\delta(\lambda - \lambda_j)d\lambda, & \lambda < 0, \end{cases}$$

$$C_j = \left(\int_0^\infty \varphi_j(x)^2 dx\right)^{-1}.$$

1.8 Partial Data and Local Uniqueness

When the potential is spherically symmetric, the three-dimensional Schrödinger equation is reduced to a set of one-dimensional equations

$$\left(-\frac{d^2}{dx^2} + \frac{\ell(\ell + 1)}{x^2} + q(x) - \lambda\right)u = 0, \quad 0 < x < \infty, \quad \ell = 0, 1, 2, \cdots.$$

Regge [168] considered an analytic continuation with respect to the angular momentum ℓ of solutions and used it in scattering theory. Loeffel [135] and Ramm [165] used this *complex angular momentum* method for the inverse scattering from a fixed energy. The following theorems from complex analysis are in the background of this idea. The first one is Carleson's theorem (see [25, p. 153]).

Theorem 1.21 *Assume that* $f(z)$ *is analytic in* $\{\text{Re}\, z > 0\}$, *and let*

$$h(\varphi) = \limsup_{r \to \infty} \frac{\log |f(re^{i\varphi})|}{r}.$$

Assume also that there exists a constant $B < \pi$ *such that*

$$h(\varphi) < B |\sin \varphi|, \quad -\pi/2 < \varphi < \pi/2.$$

Then $f(z) = 0$ *in* $\{\text{Re}\, z > 0\}$ *provided* $f(n) = 0$ *for all* $n = 0, 1, 2, \cdots$.

The *Nevanlinna class* is the set of analytic functions $f(z)$ in $\{\text{Re}\, z > 0\}$ satisfying

$$\sup_{0 < r < 1} \int_{-\pi}^{\pi} \log^+ \left| f\left(\frac{1 - re^{i\varphi}}{1 + re^{i\varphi}} \right) \right| d\varphi < \infty, \tag{1.62}$$

where $\log^+ x = \max\{\log x, 0\}$.

Theorem 1.22 *Let* $\mathscr{L}$ *be a subset of positive integers such that* $\sum_{\ell \in \mathscr{L}} 1/\ell = \infty$. *Assume that* $f(z)$ *belongs to the Nevanlinna class and* $f(\ell) = 0$ *for all* $\ell \in \mathscr{L}$. *Then,* $f(z) = 0$ *in* $\{\text{Re}\, z > 0\}$.

When $\mathscr{L} = \{0, 1, 2, \cdots\}$, this theorem is classical. It is proved in the above form in [165].

These theorems imply that the physical phase shifts δ_ℓ for $\ell = 0, 1, 2, \cdots$ have unique analytic extensions with respect to complex ℓ. Ramm's results says that to recover the potential, a part of data δ_ℓ satisfying $\sum_\ell 1/\ell = \infty$ (*Müntz type condition*) is sufficient.

Let us consider two Schrödinger operators $H_j = -\frac{d^2}{dx^2} + q_j(x)$ on $(0, \infty)$, where $q_j \in L^1_{loc}([0, \infty))$, with Dirichlet condition at $x = 0$. Letting $m_j(z)$ be their Weyl functions, Gesztesy–Simon [65] proved that if

$$m_1(z) - m_2(z) = O(e^{-2\text{Im}\,\sqrt{z}a}), \quad \text{as} \quad |z| \to \infty$$

along the ray $\arg z = \pi - \epsilon$, where $a > 0$ and $0 < \epsilon < \pi/2$, then

$$q_1(x) = q_2(x), \quad 0 < x < a.$$

This result is significant in that only the asymptotic behavior of spectral data is used and the result is local. The following uniqueness theorem for the finite Laplace transform is used in the proof (see [177]).

Theorem 1.23 *Let* $f(x) \in L^1(0, a)$ *and assume that* $\int_0^a f(x)e^{-\lambda x}dx = O(e^{-\lambda a})$ *as* $\lambda \to \infty$. *Then* $f(x) = 0$ *on* $[0, a]$.

Using these ideas, Daudé, Karman, and Nicoleau [53] considered inverse scattering for the Dirac operator on spherically symmetric asymptotically hyperbolic manifold $\Sigma = \mathbf{R}_x \times S^2_{\theta,\varphi}$ endowed with a metric $ds^2 = dx^2 + a(x)^{-2}(d\theta^2 + \sin^2\theta \, d\varphi^2)$. Assume that $a(x) = a_\pm e^{\pm\kappa x} + O(e^{3\kappa x})$ as $x \to \pm\infty$, where $\pm\kappa < 0$. Expanding by spherical harmonics, one obtains a family of S-matrices $S(\lambda, n)$, $n = 1, 2, \cdots$, each of which is a 2×2 unitary matrix with diagonal entry $T(\lambda, n)$ and off-diagonal entries $R(\lambda, n)$, $L(\lambda, n)$. Then, they proved that given two metrics $ds^2, d\tilde{s}^2$ with data $L(\lambda, n)$ and $\tilde{L}(\lambda, n)$ for a fixed energy λ, the following two assertions (1) and (2) are equivalent:

(1) $L(\lambda, n) = \tilde{L}(\lambda, n) + O(e^{-2nB})$, $\quad n \to \infty$,

(2) $\exists k \in \mathbf{Z} \; s.t. \; a(x) = \tilde{a}(x + k\pi/\lambda)$, $\quad \forall x < C(B)$,

where $C(B)$ is an explicit constant satisfying $C(B) \to \infty$ as $B \to \infty$. This theorem has a clear physical meaning. One sends incident waves from $x = -\infty$ and observes the scattered waves at $x = -\infty$. Then, these data determine the part $x < C(B)$ of the manifold even if we have only the approximate data $L(\lambda, n)$ for one fixed energy. In particular, it implies that if $L(\lambda, n) = \tilde{L}(\lambda, n)$ for a fixed $\lambda \neq 0$ and all n, then $a(x) = \tilde{a}(x + k\pi/\lambda)$ for all x. Using this theorem, they solved the inverse scattering problem for Reissner–Nordström–de Sitter metric and Kerr–Newman–de Sitter metric appearing in general relativity.

Chapter 2
Multi-Dimensional Inverse Boundary Value Problems

The inverse spectral problem for boundary value problem of differential operators has been raised and studied from various viewpoints. We recall some of the sources of the problem in this chapter.

We start with a uniqueness theorem in the inverse boundary value problem: For the Dirichlet problem, the potential is determined by the eigenvalues and boundary values of the derivatives of eigenfunctions. It is apparently an analogue of the one-dimensional results. However, there is a big difference between the inverse spectral theory in one-dimension and that in multi-dimensions. In the latter case, these spectral data are overdetermined and do not allow isospectral deformation. As in the case of the one-dimensional problem, the key tool is the D-N map, and its knowledge for an arbitrarily fixed energy is sufficient to determine the potential. This result has an important application in electric impedance tomography. Another interesting issue is the determination of Riemannian metric from spectral data. For the sake of simplicity, we mainly deal with the Dirichlet problem. However, the results can be extended to other boundary conditions.

2.1 Multi-Dimensional Borg–Levinson Theorem

In a bounded domain $\Omega \subset \mathbf{R}^n$, $n \geq 2$, with smooth boundary $\partial\Omega$,[1] consider the Dirichlet problem

$$\begin{cases} (-\Delta + V(x) - \lambda)u = 0 & \text{in} \quad \Omega, \\ u = f & \text{on} \quad \partial\Omega, \end{cases}$$

---

[1] Actually the Lipschitz continuity is enough. However, it requires a harder analysis.

© The Author(s), under exclusive licence to Springer Nature Singapore Pte Ltd. 2020
H. Isozaki, *Inverse Spectral and Scattering Theory*, SpringerBriefs
in Mathematical Physics 38, https://doi.org/10.1007/978-981-15-8199-1_2

where $V(x)$ is a real function. If the potential is absent, one can define the self-adjoint Dirichlet Laplacian $H_0 = -\Delta_D$ by

$$D(H_0) = H_0^1(\Omega) \cap H^2(\Omega), \quad H_0 u = -\Delta u, \quad u \in D(H_0).$$

Assuming that $V \in L^2(\Omega)$ for $1 \leq n \leq 3$, and $V \in L^p(\Omega)$, $p > n/2$, for $n \geq 4$, one can show that the operator $H = H_0 + V$, $D(H) = D(H_0)$, is self-adjoint, and the resolvent $(H - z)^{-1}$ is compact on $L^2(\Omega)$. Let

$$\lambda_1(V) < \lambda_2(V) \leq \cdots \leq \lambda_n(V) \leq \cdots$$

be the Dirichlet eigenvalues, and $\varphi_1(x, V), \varphi_2(x, V), \cdots$ be the associated complete orthonormal system of eigenvectors. Note that the first eigenvalue is known to be simple, i.e. of multiplicity one. Letting ν be the unit outer normal to $\partial\Omega$, we adopt

$$\frac{\partial \varphi_1}{\partial \nu}(x, V)\Big|_{\partial\Omega}, \ \frac{\partial \varphi_2}{\partial \nu}(x, V)\Big|_{\partial\Omega}, \ \cdots$$

as the spectral data, where $u|_{\partial\Omega}$ means the restriction of u to $\partial\Omega$.

Nachman et al. [151] proved that, under some regularity assumption on the potential, these spectral data determine the potential $V(x)$ uniquely. It is a multi-dimensional analogue of the one-dimensional Borg–Levinson theorem. See also [37]. The following remark given by Isozaki [85] explains the difference between the one-dimensional and multi-dimensional problems. The Dirichlet problem

$$\begin{cases} (-\Delta + V(x) - \lambda)u = 0 & \text{in} \quad \Omega, \\ u = f & \text{on} \quad \partial\Omega, \end{cases} \tag{2.1}$$

is solvable for $\lambda \notin \sigma(H)$. We define the D-N map $\Lambda(\lambda; V)$ by

$$\Lambda(\lambda; V)f = \frac{\partial u}{\partial \nu}\Big|_{\partial\Omega},$$

where u is a solution to (2.1). It is a bounded operator from $H^{3/2}(\partial\Omega)$ to $H^{1/2}(\partial\Omega)$. By using the quadratic form technique, one can show that $\Lambda(\lambda; V)$ is bounded from $H^{1/2}(\partial\Omega)$ to $H^{-1/2}(\partial\Omega)$, and is a meromorphic function of $\lambda \in \mathbf{C}$ with poles at $\sigma(H)$. We put

$$\varphi_{\lambda,\omega}(x) = e^{i\sqrt{\lambda}\omega \cdot x}, \quad \lambda \in \mathbf{C} \setminus (-\infty, 0), \quad \omega \in S^{n-1},$$

and let $\langle \ , \ \rangle$ and $(\ , \ )$ be the inner products of $L^2(\partial\Omega)$ and $L^2(\Omega)$, respectively. Letting $R(\lambda)$ be the resolvent of $-\Delta + V(x)$ for the above Dirichlet problem, we can prove the following formula.

Lemma 2.1 *We put*

$$S(\lambda, \theta, \omega; V) = \langle \Lambda(\lambda, V)\varphi(\lambda, \omega), \overline{\varphi_{\lambda, -\theta}} \rangle.$$

Then we have the following expression:[2]

$$S(\lambda, \theta, \omega; V) = -\frac{\lambda}{2}(\theta - \omega)^2 \int_{\Omega} e^{-i\sqrt{\lambda}(\theta - \omega) \cdot x} dx$$

$$+ \int_{\Omega} e^{-i\sqrt{\lambda}(\theta - \omega) \cdot x} V(x) dx - (R(\lambda)V\varphi_{\lambda, \omega}, \overline{V\varphi_{\lambda, -\theta}}). \tag{2.2}$$

An interesting feature is that this formula is similar to Heisenberg's S-matrix for $H = -\Delta + V(x)$ on $\mathbf{R}^n$, where $V(x)$ is extended to be 0 outside of Ω. In fact, if we replace the first term of the right-hand side of (2.2) by $\delta(\theta - \omega)$ (the δ-function on the sphere) and $R(\lambda)$ by the resolvent of $H = -\Delta + V(x)$ in $\mathbf{R}^n$, we obtain the S-matrix (see (3.3) and (3.4)).

We use a variant of Born approximation often utilized in scattering theory. Given $0 \neq \xi \in \mathbf{R}^n$, choose $\eta \in S^{n-1}$ such that η is orthogonal to ξ. For a large parameter ℓ, put

$$\begin{cases} \theta_\ell = c_\ell \eta + \xi/2\ell, \quad c_\ell = \left(1 - |\xi|^2/(4\ell^2)\right)^{1/2}, \\ \omega_\ell = c_\ell \eta - \xi/2\ell, \\ \sqrt{t_\ell} = \ell + i. \end{cases}$$

Then, we have

$$\lim_{\ell \to \infty} S(t_\ell, \theta_\ell, \omega_\ell; V) = -\frac{|\xi|^2}{2} \int_{\Omega} e^{-ix \cdot \xi} dx + \int_{\Omega} e^{-ix \cdot \xi} V(x) dx.$$

Therefore, if one knows the D-N map $\Lambda(\lambda; V)$ for all $\lambda \in \mathbf{C} \setminus (-\infty, 0)$, one can reconstruct $V(x)$.

By a simple computation, one can show that $\Lambda(\lambda; V)$ has, formally, the integral kernel[3]

$$\sum_{i=1}^{\infty} \frac{1}{\lambda - \lambda_i(V)} \frac{\partial \varphi_i}{\partial \nu}(x, V) \frac{\partial \varphi_i}{\partial \nu}(y, V), \quad x, y \in \partial\Omega, \tag{2.3}$$

---

[2]The proof relies on a simple integration by parts.

[3]Take $\widetilde{f} \in H^2(\Omega)$ such that $\widetilde{f} = f$ on $\partial\Omega$, and seek a solution of (2.1) in the form $u = \widetilde{f} + v$. Then, $(-\Delta + V - \lambda)v = (\Delta - V + \lambda)\widetilde{f} =: F$. Hence, $v = R(\lambda)F = \sum_{i=1}^{\infty}(\lambda_i - \lambda)^{-1}(F, \varphi_i)\varphi_i$. Integration by parts then yields $u = \sum_{i=1}^{\infty}(\lambda - \lambda_i)^{-1}\langle f(y), \frac{\partial \varphi_i(y)}{\partial \nu}\rangle \varphi_i(x)$.

(actually, this series is divergent). Now, assume that for two potentials V_1 and V_2, their eigenvalues and normal derivatives of eigenvectors coincide except for a finite number of i's. Then, there exists N such that the terms in (2.3) with $i \geq N$ coincide for V_1 and V_2. Hence, the difference of their D-N maps will consist of a finite sum of low energy eigenvalues and eigenvectors, and will vanish as $\lambda \to \infty$. This observation can be justified, and we have the following theorem ([85, 152]).

Theorem 2.1 *Assume that* $V_1, V_2 \in C^\infty(\overline{\Omega})$ *and there exists* $N \geq 1$ *such that*

$$\lambda_i(V_1) = \lambda_i(V_2), \quad \frac{\partial \varphi_i}{\partial v}(x, V_1)\Big|_{\partial\Omega} = \frac{\partial \varphi_i}{\partial v}(x, V_2)\Big|_{\partial\Omega}$$

for all $i \geq N$. *Then,* $V_1 = V_2$.

This theorem asserts that V is determined only from high energy eigenvalues and eigenvectors. Therefore, unlike the one-dimensional case, we cannot expect the isospectral deformation for Schrödinger operators $-\Delta + V(x)$ in multi-dimensions. This fact can be extended to the case of variable coefficients $-\sum_{i,j=1}^{n} \partial_i (a_{ij}(x)\partial_j) + V(x)$, when $a_{ij}(x)$ is sufficiently close to δ_{ij}. This theorem, often called inverse problem with incomplete spectral data, has been extended to cases where spectral data asymptotically coincide, or for singular potentials, higher order operators and cylindrical domains [40, 107, 122, 156, 164]. For the case of asymptotically equal data, the following theorem is proven in [112],

Theorem 2.2 *Assume that*

$$\lim_{i\to\infty} \left|\lambda_i(V_1) - \lambda_i(V_2)\right| = 0, \quad \sum_{i=1}^{\infty} \left\|\frac{\partial}{\partial v}\varphi_i(V_1) - \frac{\partial}{\partial v}\varphi_i(V_2)\right\|_{L^2(\partial\Omega)} < \infty.$$

Then, $V_1 = V_2$.

2.2 Gel'fand Problem

In the international congress of mathematics held in Amsterdam in 1954, I. M. Gel'fand gave a survey talk on the development of functional analysis [63]. One of the topics is on differential operators, and centered around boundary value problems, spectral theory, eigenfunction expansion and inverse problems, all of which are related with quantum mechanics and quantum field theory. For the one-dimensional case, the results had been established by the works of Krein, Marchenko, Gel'fand-Levitan. The problems were naturally extended to multi-dimensions: inverse boundary value problems and inverse scattering problems. In particular, he raised the following question. Consider the boundary value problem

$$\begin{cases} (-\Delta + V)u = \lambda u, & in \quad \Omega, \\ \dfrac{\partial u}{\partial \nu} = 0, & on \quad \partial \Omega \end{cases}$$

in an exterior domain Ω. Let $E(x, y; \lambda)$ be kernel of the spectral decomposition of $-\Delta + V$. Then, does $E(x, y; \lambda)\big|_{x,y \in \partial \Omega}$ determine $V(x)$?

Belishev and Kurylev [16] extended this question for compact manifolds. Let M be an m-dimensional Riemannian manifold with boundary $S = \partial M$, equipped with Riemannian metric $g_{jk} dx^j dx^k$. Let $g = \det(g_{jk})$ and consider the differential operator

$$A = -\frac{1}{\sqrt{g}} (\partial_j + ib_j)\sqrt{g} g^{jk} \mu (\partial_k + ib_k) + q \quad in \quad L^2(M, \sqrt{g}dx),$$

where $(g^{jk}) = (g_{jk})^{-1}$, with boundary condition

$$\left(\partial_\nu + ib_\nu + \sigma\right)u\big|_S = 0 \quad on \quad S,$$

where $b = (b_1, \cdots, b_m)$, μ, q, σ are real-valued and $\mu \geq \mu_0$ for a constant $\mu_0 > 0$, ∂_ν is the normal derivative, and b_ν is the normal component of b at S. In the application to electromagnetism, μ is a conductivity, σ is a boundary impedance. In quantum mechanics, b and q are electric and magnetic potentials. The operator A has the discrete spectrum $\lambda_1 \leq \lambda_2 \leq \cdots \to \infty$. Let $\{\varphi_k(x)\}$ be the associated orthonormal eigenfunctions. Let us call the triple $(S, \{\lambda_k\}, \{\varphi_k|_S\})$ the *boundary spectral data (BSD)* of A, and $(M, (g_{jk}), A)$ the operator system. The generalized question is then as follows:

Generalized Gel'fand problem: Does BSD $(S, \{\lambda_k\}, \{\varphi_k|_S\})$ determine the manifold M and the operator A?

Belishev and Kurylev [16] solved this problem by the *boundary control method (BC- method)* initiated by Belishev [15], which is closely related with the Gel'fand–Levitan theory and will be explained later.

By the gauge transformation $A \to e^{ic} A e^{-ic}$, the magnetic field b_k is transformed to $b_k + \partial_k c$. It is natural to allow the generalized gauge transformation as follows:

Definition 2.1 Two BSDs $(S, \{\lambda_k\}, \{\varphi_k|_S\})$ and $(R, \{\mu_k\}, \{\psi_k|_R\})$ of the operator systems (M, g, A), (N, h, B) are said to be equivalent if

(1) $\lambda_k = \mu_k, \ \forall k \geq 1$,
(2) there exists a conformal diffeomorphism $\phi : S \to R$ and a real-valued function $\kappa_0 > 0$ on S such that

$$\phi^*(\psi_k|_R) = \kappa_0 \varphi_k|_S, \qquad \phi^*(h|_R) = \kappa_0^{4/m} g|_S.$$

Lemma 2.2 *The BSDs $\left(S, \{\lambda_k\}, \{\varphi_k|_S\}\right)$ and $\left(R, \{\mu_k\}, \{\psi_k|_R\}\right)$ of the operator systems $\left(M, g, A\right)$ and $\left(N, h, B\right)$ are equivalent if and only if there exists a conformal diffeomorphism $\Phi : M \to N$ and a complex-valued function $\kappa \neq 0$ such that*

$$B = \Phi \circ \left(\kappa A \kappa^{-1}\right) \circ \Phi^{-1}, \quad \Phi|_S = \phi, \quad \kappa|_S = \kappa_0.$$

The generalized Gel'fand problem is solved in the following way.

Theorem 2.3 *For a BSD $\left(S, \{\lambda_k\}, \{\varphi_k|_S\}\right)$, there exists a unique operator system $\left(M, g, A\right)$, up to a gauge transformation.*

The incomplete spectral data problem is also extended to compact manifolds [103].

2.3 Kac Problem

In 1966, M. Kac [100] raised the following problem with an impressive title "Can one hear the shape of a drum?" and an interesting episode of D. Hilbert and H. Weyl:

What can we know about the geometry of a Riemannian manifold from the knowledge of eigenvalues of its Laplace operator?

Suppose we are given two compact Riemannian manifolds whose all eigenvalues of Laplace operators coincide. Are they isometric? The answer is negative. There is a counterexample of 16-dimensional tori due to Milnor. Then, the next issue is to extract geometric properties as much as possible from the knowledge of eigenvalues. Kac computed the asymptotic expansion of the trace of the heat kernel of a planar domain and showed that one can know the area of the domain, the length of the boundary, and in the case of polygonal region the number of holes in the domain. Kac's paper stimulated two directions of research, isospectral manifolds and spectral invariants.

2.3.1 Isospectral Manifolds

These problems attracted so many people that an extensive literature has been devoted to them. Urakawa [190] constructed non-congruent regions in Euclidean space with the same Dirichlet (and Neumann) eigenvalues. Non-Euclidean case had been studied for lens space by Yamamoto and Ikeda [80] and Ikeda [79]. Sunada [183] developed a general method of constructing isospectral manifolds. An exposition of Sunada's theory is given by Bérard [21]. For more details of this subject, see a review article of Urakawa [191]. Below, let us give an example of two

Fig. 2.1 Isospectral
manifolds

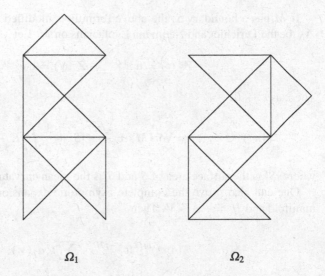

$$\Omega_1 \qquad\qquad\qquad\qquad \Omega_2$$

non-congruent regions having the same Dirichlet and Neumann eigenvalues due to
Chapmann [38] (Fig. 2.1).

2.3.2 Spectral Invariants

Spectral invariants are the numbers, formulas, etc., determined only by the spectrum.
The trace of the heat kernel is often employed to compute them. Let M be a compact
m-dimensional Riemannian manifold without boundary, and put

$$Z(t) = \operatorname{tr} e^{t\Delta} = \sum_{j=1}^{\infty} e^{-t\lambda_j},$$

where $\lambda_1 \le \lambda_2 \le \cdots \to \infty$ are eigenvalues of the Laplace operator $-\Delta$ on M.
Mckean and Singer [141] proved that as $t \to 0$,

$$(4\pi t)^{m/2} Z(t) = \operatorname{Vol}(M) + \frac{t}{3}\int_M K + \frac{t^2}{180}\int_M (10A - B + 2C) + \cdots,$$

where K is the scalar curvature and A, B, C are polynomials of the curvature tensor.
If $m = 2$, this formula reads

$$Z(t) = \frac{|M|}{4\pi t} + \frac{E}{t} + \frac{\pi t}{60}\int_M K^2 + \cdots,$$

where $|M|$ is the area of M, $E = \frac{1}{2\pi}\int_M K$ is the Euler characteristic of M.

If M has a boundary S, the above formula is modified as follows. Let Δ_D and Δ_N be the Dirichlet and Neumann Laplacians on M. Let

$$Z_+(t) = \operatorname{tr} e^{t\Delta_N}, \quad Z_-(t) = \operatorname{tr} e^{t\Delta_D}.$$

Then, as $t \to 0$,

$$(4\pi t)^{m/2} Z_\pm(t) = \operatorname{Vol}(M) \pm \frac{\sqrt{\pi t}}{2}|S| + \frac{1}{3}\int_M K - \frac{1}{6}\int_S J + \cdots,$$

where $|S|$ is the surface area of S and J is the mean curvature.

One can also prove the complete asymptotic expansion. Let M be a compact manifold and $H = -\Delta + V$. Then

$$(4\pi t)^{m/2} \operatorname{tr} e^{-tH} \sim \sum_{j=0}^{\infty} t^j a_j(V),$$

where $a_j(V)$'s are integrals of certain functions over M which are universal polynomials in the covariant derivatives of V and curvature tensors of M.

Using this asymptotic expansion, one can prove (see, e.g., Brüning [28], Brooks et al. [27]) that the set of Riemannian metrics on M isospectral to (M, g) is compact with respect to C^∞-topology on Riemannian metrics. See also a survey of Perry [157].

2.4   Calderón Problem

2.4.1   *Complex Geometrical Optics Solutions*

The Calderón problem is another source of inverse problems, and has been making a rapid progress. We stop at this issue, and look at the relations to other topics. For an excellent survey, see [189] and the references therein.

It starts from the paper of Calderón [33]. Although the paper was presented in 1980, Calderón had been considering this problem from the 1950s when he was working as an engineer in the state oil company of Argentine. The original problem is practical and simple. Given a bounded body $\Omega \subset \mathbf{R}^n$, compute its electric conductivity from the measurement of electric current and voltage on the boundary $\partial\Omega$. This is also called the problem of *electric impedance tomography* (EIT). Mathematically, this problem is formulated as follows. The electric conductivity is represented by a positive function $\gamma(x)$ on Ω. In the absence of sinks or sources of current, the voltage potential $u(x)$ satisfies the equation

$$\begin{cases} \nabla \cdot (\gamma(x)\nabla u) = 0 & \text{in} \quad \Omega, \\ u = f & \text{on} \quad \partial\Omega. \end{cases} \tag{2.4}$$

The mapping from voltage to current on $\partial\Omega$ is the D-N map:

$$\Lambda_\gamma f = \gamma \frac{\partial u}{\partial \nu} \quad \text{on} \quad \partial\Omega, \tag{2.5}$$

where ν is the outer unit normal to $\partial\Omega$. The problem is: Compute γ from the D-N map.[4] Calderón considered the linearization of the map: $\gamma \to \int_{\partial\Omega}(\Lambda_\gamma f)\overline{f}dS = \int_\Omega \gamma |\nabla u|^2 dx$, and proved its injectivity. His main idea is to use the harmonic function $u = e^{ix\cdot\zeta}$, where $\zeta \in \mathbf{C}^n$ satisfies $\zeta \cdot \zeta = 0$. Letting $\zeta = k + i\ell, \zeta \cdot \zeta = 0$ implies $|k| = |\ell|, k \cdot \ell = 0$. Therefore, $e^{i\zeta \cdot x} = e^{-\ell x}e^{ik\cdot x}$ is exponentially growing in the half-space $-\ell \cdot x > 0$ and exponentially decaying in the opposite half-space $-\ell \cdot x < 0$. In the later development of mathematics for EIT, this solution is called *exponentially growing solution*. It is also called *complex geometrical optics solution* and abbreviated as *CGO* solution.

Sylvester and Uhlmann [184] made this CGO solution to be a main tool in inverse boundary value problems. Let us look at their idea briefly. Starting from the identity

$$\gamma^{-1/2}(\nabla \cdot \gamma\nabla)(\gamma^{-1/2}u) = (\Delta - q(x))u, \quad q(x) = \frac{\Delta\sqrt{\gamma}}{\sqrt{\gamma}}, \tag{2.6}$$

the problem is reduced to that for the Schrödinger operator $\left(- \Delta + q(x)\right)u = 0$. They sought the solution in the form $u = e^{i\zeta \cdot x}(1 + \psi)$, which is rewritten as

$$(-\Delta - 2i\zeta \cdot \nabla + q)\psi = -q.$$

This suggests that the Green operator $(-\Delta - 2i\zeta \cdot \nabla)^{-1}$ plays a key role. It is interesting that the multi-dimensional inverse boundary value problem and the inverse scattering problem were developed from this same idea, *independently*. The founders did not know each other. The exponentially growing solution of the Schrödinger equation had already appeared in the 1960s in Faddeev's theory of inverse scattering [59]. Consider the Schrödinger equation $(-\Delta + V(x) - E)\varphi = 0$. Take $\zeta = (\zeta_1, \cdots, \zeta_n) \in \mathbf{C}^n$ such that $\zeta \cdot \zeta = \sum_{j=1}^{n} \zeta_j^2 = E$ and let $\varphi = e^{i\zeta \cdot x}(1 + v)$. Then v satisfies $(-\Delta - 2i\zeta \cdot \nabla + V)v = -V(x)$. This leads us to the equation

$$(-\Delta - 2i\zeta \cdot \nabla)u = f.$$

---

[4]Physically, it is more natural to consider N-D map, which assigns current to voltage. In some sense, this is also true mathematically, since N-D map is bounded while D-N map is unbounded.

A natural choice of the solution is the one written by the Fourier transformation

$$u(x) = (2\pi)^{-n/2} \int_{\mathbf{R}^n} \frac{e^{ix\cdot\xi}}{\xi^2 + 2\zeta\cdot\xi} \widehat{f}(\xi)d\xi =: \widetilde{G}(\zeta)f. \tag{2.7}$$

This is the Green operator introduced by Faddeev. As is seen above, the essential idea is to consider the solution of the form $e^{i\zeta\cdot x}$ of the Helmholtz equation $(-\Delta - E)u = 0$ with complex phase function $\zeta\cdot x$.

For $s \in \mathbf{R}$, we introduce a function space $L^{2,s}$ by

$$f \in L^{2,s} \iff \|f\|_s = \left(\int_{\mathbf{R}^n} (1 + |x|)^{2s} |f(x)|^2 dx \right)^{1/2} < \infty. \tag{2.8}$$

The following estimate is due to Sylvester and Uhlmann [184]. See also [72].

Theorem 2.4 *Let* $-1 < s < 0$ *and* $E, T > 0$. *Then there exists a constant* $C > 0$ *such that for any* $\zeta \in \mathbf{C}^n$ *satisfying* $\zeta\cdot\zeta = E$, *and* $|\mathrm{Im}\,\zeta| > T$, *the following inequality holds*

$$\|\widetilde{G}(\zeta)f\|_s \le \frac{C}{|\zeta|} \|f\|_{s+1}.$$

Let us touch upon the proof of Theorem 2.4. Let

$$F(x, y) = \frac{1}{\pi(x + iy)}.$$

Then, we have

$$\frac{1}{2}\left(\frac{\partial}{\partial x} + i\frac{\partial}{\partial y}\right) F(x, y) = \delta.$$

Namely, $F(x, y)$ is a fundamental solution to the $\bar{\partial}$-operator, and the solution of the $\bar{\partial}$-equation $\bar{\partial}u = f$ is written as

$$u(x, y) = \frac{1}{\pi} \int_{\mathbf{R}^2} \frac{f(x', y')}{x - x' + i(y - y')} dx'dy' = \frac{1}{\pi i} \int_{\mathbf{R}^2} e^{i(x\xi + y\eta)} \frac{\widehat{f}(\xi, \eta)}{\xi + i\eta} d\xi d\eta.$$

Then, the convolution operator $F * f$ is bounded from $L^{2,s-1}(\mathbf{R}^2)$ to $L^{2,s}(\mathbf{R}^2)$ for $0 < s < 1$. When $\zeta \in \mathbf{C}^n$ satisfies $\zeta\cdot\zeta = E$, letting $\zeta = k + i\ell$, we have $k^2 - \ell^2 = E, k\cdot\ell = 0$. Therefore, the denominator of (2.7) is rewritten as $\xi^2 + 2\zeta\cdot\xi = (\xi + k)^2 - k^2 + 2i\ell\cdot\xi$, whose zeros form an $(n - 2)$-dimensional sphere. By localizing near this sphere, one can use the above estimate to prove Theorem 2.4. Another proof based on ordinary differential equation is given in [176], in which the theorem is extended to product manifolds $\mathbf{R} \times M$ where M is a compact manifold.

Actually, Sylvester and Uhlmann considered the case $E = 0$, however the case $E > 0$ can be dealt with by the same method. Furthermore, one can also take the limit Im $\zeta \to 0$ of the above estimate (with different weight s and $s + 1$), which is useful in the application to scattering theory, and will be discussed in Chap. 3.

Sylvester and Uhlmann [184] then solved the inverse boundary value problem in the following way.

Theorem 2.5 *Let $n \geq 3$, and assume $\gamma_i \in C^2(\overline{\Omega})$, $i = 1, 2$. If $\Lambda_{\gamma_1} = \Lambda_{\gamma_2}$, then $\gamma_1 = \gamma_2$ in Ω.*

Note that Theorem 2.5 also implies the multi-dimensional Borg–Levinson theorem (Theorem 2.1).

Let us explain the outline of the proof of Theorem 2.5. As was done in (2.6), the problem is reduced to that for the Schrödinger equation. Given two potentials q_i, $i = 1, 2$, note the formula

$$\int_\Omega (q_1 - q_2)u_1 u_2 dx = \int_{\partial\Omega} \left(\frac{\partial u_1}{\partial \nu} u_2 - u_1 \frac{\partial u_2}{\partial \nu} \right) dS, \qquad (2.9)$$

for u_i satisfying $(-\Delta + q_i)u_i = 0$ on Ω, $i = 1, 2$. Take two $\zeta_i \in \mathbf{C}^n$ satisfying $\zeta_i \cdot \zeta_i = 0$. Then, using Theorem 2.4, one can construct u_i of the form

$$u_i = e^{ix \cdot \zeta_i} (1 + \psi_i(x, \zeta_i)),$$

where $\|\psi_i\|_{L^2(\Omega)} \leq C/|\zeta_i|$. Given $\xi \in \mathbf{R}^n$, take $\eta, \ell \in \mathbf{R}^n$ such that

$$\eta \cdot \xi = \xi \cdot \ell = \eta \cdot \ell = 0, \quad |\eta|^2 = |\xi|^2 + |\ell|^2, \qquad (2.10)$$

and put

$$i\zeta_1 = \frac{\eta}{2} + i\frac{-\xi + \ell}{2}, \quad i\zeta_2 = -\frac{\eta}{2} + i\frac{-\xi - \ell}{2}.$$

We then have $u_1 u_2 \sim e^{ix \cdot \zeta_1} e^{ix \cdot \zeta_2} = e^{-ix \cdot \xi}$, since $\zeta_1 + \zeta_2 = \xi$. Then, (2.9) turns out to be (extending $q_i(x)$ to be 0 outside Ω)

$$\widehat{q_1}(\xi) - \widehat{q_2}(\xi) = -\int_\Omega e^{-ix \cdot \xi} (q_1 - q_2)(\psi_1 + \psi_2 + \psi_1 \psi_2) dx.$$

Letting $|\zeta_i| \to \infty$, we obtain $q_1(x) = q_2(x)$.

Note that (2.10) is possible only when $n \geq 3$.

The stability for the reconstruction is an important issue, and let us cite here[5], although there are plenty of stability results.

2.4.2   Anisotropic Conductivity

In the above arguments, we have assumed that $\gamma(x)$ in (2.4) is a scalar function, however, this is not sufficient for the practical application. For example, in the case of medical application, the electric conductivity for the muscle is known to be strongly anisotropic. In such a case, $\gamma(x)$ is a positive definite symmetric matrix $(\gamma_{ij}(x))$, and the D-N map in (2.5) is defined by $\Lambda_\gamma f = \sum_{i,j=1}^{n} \nu_i \gamma_{ij} \frac{\partial u}{\partial x_j}$ on $\partial\Omega$. We must then take notice of the following fact. If $\Phi : \Omega \to \Omega$ is a diffeomorphism, by making a change of variables $y = \Phi(x)$ and putting $v(y) = u(\Phi(y))$, we have $\nabla \cdot (\Phi_* \gamma) \nabla v = 0$ in Ω. Here, letting $d\Phi(x)$ be the differential of $\Phi(x)$ and $J\Phi(x)$ its determinant, $(\Phi_* \gamma)(y) = \frac{1}{J\Phi(x)} d\Phi(x)\gamma(x)d\Phi^t(x)\Big|_{x=\Phi^{-1}(y)}$. Moreover, if $\Phi\big|_{\partial\Omega}$ is the identity, we have $\Lambda_{\Phi_*\gamma} = \Lambda_\gamma$. This means that the D-N map is invariant under the diffeomorphism leaving the boundary invariant. Then, the natural question is whether or not $\gamma(x)$ is determined uniquely from the D-N map up to a diffeomorphism leaving the boundary invariant. The same problem can be formulated for the inverse boundary value problem on a domain in a Riemannian manifold.

In dimension $n \geq 3$, for the case of piecewise real analytic conductivities, this problem was solved affirmatively by Kohn and Vogelius [115] and Lassas and Uhlmann [126]. For the C^∞-case, it is still making progress, which we discuss in the next section. However, in 2-dimensions we have satisfactory results, which will be summarized in Sect. 2.4.4.

A remarkable non-uniqueness example for anisotropic conductivities is obtained by Greenleaf et al. [69]. It was refound independently in engineering and opened a new area of research in mathematics and material science called invisibility or cloaking. See the survey work of Uhlmann [189].

2.4.3   Carleman Estimates

Let Ω be a bounded domain in $\mathbf{R}^n$ and P a differential operator on Ω. The Carleman estimate is, in its simplest form, the following inequality,

$$s^\alpha \|e^{s\varphi}u\|_{L^2(\Omega)} \leq C\|e^{s\varphi}Pu\|_{L^2(\Omega)}, \quad \forall u \in C_0^\infty(\Omega) \tag{2.11}$$

where $\alpha > 0$, the constant C does not depend on a large parameter $s > 0$ and φ is a real function satisfying $\nabla\varphi \neq 0$ on Ω. Carleman derived this inequality to prove the unique continuation property. Namely, for a domain $\Omega \subset \mathbf{R}^2$, given $q(x) \in L^\infty(\Omega)$ and an open subset $\omega \subset \Omega$, if u satisfies

$$(-\Delta + q(x))u = 0 \quad \text{in} \quad \Omega,$$

and $u = 0$ on ω, then $u = 0$ on the whole Ω [34]. Carleman's method is not restricted within the classical classification of types of differential equations (elliptic, parabolic, hyperbolic), and became a powerful tool to prove the uniqueness of the Cauchy problem for general partial differential equations. Calderón [32] opened a way to Carleman estimates via calculus of pseudo-differential operators, giving conditions in terms of characteristic roots. Applications of Carleman estimates to inverse problems were developed by Bukhgeim and Klibanov [31], who established a method for the unique determination of coefficients of PDE from certain data of solutions (see [113] for the full proof). Since then, Carleman estimates are regarded as one of the main streams of uniqueness and stability for the inverse problems. See Bellassoued–Yamamoto [19] for a survey.

Carleman estimates are also used by Kenig et al. [109] to construct a new type of CGO solution to the Schrödinger equation. We employ semiclassical pseudo-differential calculus. Given a function $p(x, \xi)$ on $\mathbf{R}^n \times \mathbf{R}^n$, one can associate a pseudo-differential operator, abbreviated as ΨDO,

$$p_h(x, D_x)u(x) = (2\pi)^{-n} \iint_{\mathbf{R}^n \times \mathbf{R}^n} e^{i(x-y)\cdot\xi} a(x, h\xi)u(y)dyd\xi,$$

where $h > 0$ is a small parameter. Physically, h corresponds to Planck's constant. In the regime of asymptotic expansion as $h \to 0$, there is a correspondence between the algebra of classical observables $p(x, \xi)$ and that of quantum observables $p_h(x, D_x)$. Sometimes, it is more convenient to associate the operator

$$p_h(x, D_x)u(x) = (2\pi)^{-n} \iint_{\mathbf{R}^n \times \mathbf{R}^n} e^{i(x-y)\cdot\xi} a\left(\frac{x+y}{2}, h\xi\right)u(y)dyd\xi,$$

which is usually called the Weyl quantization of $p(x, \xi)$. One of the features of Weyl ΨDO is that for a real symbol $p(x, \xi)$, $p_h(x, D_x)$ becomes a symmetric operator. See, e.g., [70], or [201].

Now let us note that (2.11) is equivalent to

$$s^\alpha \|u\|_{L^2(\Omega)} \leq C \|e^{s\varphi} P e^{-s\varphi} u\|_{L^2(\Omega)}, \quad \forall u \in C_0^\infty(\Omega). \tag{2.12}$$

In the above semiclassical approach, the parameter h is $1/s$ and the symbol of the operator $e^{s\varphi} P e^{-s\varphi}$ is $p(x, \xi + i\nabla\varphi(x))$ (up to a small term in h). We split it into

$$p(x, \xi + i\nabla\varphi(x)) = a(x, \xi) + ib(x, \xi),$$

and let A_h, B_h be the semiclassical ΨDO with symbol $a(x, \xi)$, $b(x, \xi)$. Then, we have

$$\|(A_h + iB_h)u\|^2 = \|A_h u\|^2 + \|B_h u\|^2 + (i[A_h, B_h]u, u). \tag{2.13}$$

Now, up to a small term with respect to h, the commutator $i[A_h, B_h]$ is a ΨDO with symbol $\{a, b\}$, where $\{a, b\}$ is the Poisson bracket of $a(x, \xi)$ and $b(x, \xi)$. If we have a positivity $i[A, B] \geq 0$ (up to small terms in h), we are very close to Carleman estimates. This positivity is known to hold if we have

$$a(x, \xi) = b(x, \xi) = 0 \Longrightarrow \{a(x, \xi), b(x, \xi)\} \geq 0.$$

Such an idea has been used in PDE. In the application to the EIT problem, recalling the proof of Theorem 2.5, we need to use this Carleman estimate for both of φ and $-\varphi$. In this way, we arrive at the property

$$a(x, \xi) = b(x, \xi) = 0 \Longrightarrow \{a(x, \xi), b(x, \xi)\} = 0.$$

Kenig et al. [109] called φ having this property a *limiting Carleman weight*. This work enlarged considerably the class of weights applicable to inverse boundary value problems.

One more fact is needed from functional analysis. Suppose we are given a Hilbert space X and a Banach space Y such that $Y \subset X$, Y is dense in X and $\|u\|_X \leq \|u\|_Y$ for any $u \in Y$. By Riesz' theorem, X^* is identified with X, and we have the inclusion relation $Y \subset X = X^* \subset Y^*$. We are also given a linear operator A with domain D, which is assumed to be dense in Y, such that A is closable, and, letting Z be the domain of $\overline{A}$ = the closure of A, we have

$$D \subset Z \subset Y \subset X = X^* \subset Y^* \subset Z^*.$$

By the closed graph theorem, $\overline{A} : Z \to X$ is bounded, hence $\overline{A}^* : X^* \to Z^*$ is also bounded. Assume that there exists a constant $C > 0$ such that

$$\|u\|_{X^*} \leq C\|\overline{A}^* u\|_{Z^*}, \quad \forall u \in X^*.$$

Then, by Banach's closed range theorem (see, e.g., [198]), $\overline{A}$ is onto, i.e. Ran $(\overline{A}) = X$.

Combining these ideas together, one can use limiting Carleman weights φ to solve the equation

$$e^{-\tau\varphi}(-\Delta + q)(e^{\tau\varphi}u) = f,$$

and argue as in the proof of Theorem 2.5. This makes it possible to derive the partial data results. Let $\Omega \subset \mathbf{R}^n$ be a bounded domain, and ch (Ω) its convex hull. For $x_0 \in \mathbf{R}^n \setminus \mathrm{ch}\,(\overline{\Omega})$, define

$$F(x_0) = \{x \in \partial\Omega \,;\, (x - x_0) \cdot \nu(x) \leq 0\},$$

$$B(x_0) = \{x \in \partial\Omega \,;\, (x - x_0) \cdot \nu(x) \geq 0\}.$$

For two open sets $\Gamma_D, \Gamma_N \subset \partial\Omega$, we put

$$C_q^{\Gamma_d,\Gamma_N} = \left\{ (u|_{\Gamma_D}, \partial_\nu u|_{\Gamma_N}) ; (-\Delta + q)u = 0 \text{ in } \Omega, \text{ supp}(u|_{\partial\Omega}) \subset \Gamma_D \right\}.$$

Then, we have [109]

Theorem 2.6 *Let $q_1, q_2 \in L^\infty(\Omega)$, and assume that $F(x_0) \subset \Gamma_D$, $B(x_0) \subset \Gamma_N$. If $C_{q_1}^{\Gamma_D,\Gamma_N} = C_{q_2}^{\Gamma_D,\Gamma_N}$, then, $q_1 = q_2$.*

In $\mathbf{R}^2$, any harmonic function with non-vanishing gradient is a limiting Carleman weight, while for $\mathbf{R}^n$, $n \geq 3$, the limiting Carleman weights are reduced to the following forms up to translation and constant multiple:

$$x \cdot \xi, \quad \arg x \cdot (\omega_1 + i\omega_2), \quad \log|x|, \quad \frac{x \cdot \xi}{|x|^2}, \quad \arg\left(e^{i\theta}(x + i\xi)^2\right), \quad \log\left(\frac{|x + \xi|}{|x - \xi|}\right)^2,$$

where $\omega_1, \omega_2 \in S^{n-1}, \omega_1 \cdot \omega_2 = 0, \theta \in [0, 2\pi)$, and $\xi \in \mathbf{R}^n \setminus \{0\}$. This fact suggests that the existence of limiting Carleman weight gives a certain restriction to the geometry on manifolds. In fact, Dos Santos Ferraira, C. E. Kenig, M. Salo, and G. Uhlmann proved [55] the following theorem.

Theorem 2.7 *Let (M, g) be a Riemannian manifold, and assume that there exists a limiting Carleman weight φ on M. Then, near any point of M, there are local coordinates $x = (x_1, x')$ such that in these local coordinates $\varphi(x) = x_1$ and*

$$g(x) = c(x) \begin{pmatrix} 1 & 0 \\ 0 & g_0(x') \end{pmatrix},$$

where $c(x) > 0$ and $g_0(x')$ is a metric on $\mathbf{R}^{n-1}$. Conversely, any metric of this form has the limiting Carleman weight $\varphi(x) = x_1$.

2.4.4 Two-Dimensional Problem

The boundary of a bounded domain $\Omega \subset \mathbf{R}^n$ is an $(n-1)$-dimensional manifold, hence the formal integral kernel (distribution kernel) of the D-N map Λ_γ has $2(n-1)$ number of variables. Therefore, the inverse conductivity problem : $\Lambda_\gamma \to \gamma$ seeks a function of n variables from a function of $2(n-1)$ variables. If $n \geq 3$, we have $2(n-1) - n \geq 1$. Then, we have more freedom of data than that of the target γ. In such a case, we often say that the problem is overdetermined. In the proof of Theorem 2.5, this overdeterminacy was used to compute γ from Λ_γ. For $n = 2$, the inverse problem is no longer overdetermined. This requires us to look for different ideas for the inverse problem, and the breakthrough emerged from complex analysis.

Nachman [149] used $\bar{\partial}$-equations to show the uniqueness as well as the reconstruction procedure for $\gamma(x)$ from the D-N map, which we will explain in Sect. 3.10. Here, we note that Nachman's work makes it possible to reconstruct the Riemannian metric $g_{ij}(x)$ from the D-N map, since in 2-dimensions Riemannian metrics have the form $E(x)\big((dx_1)^2 + (dx_2)^2\big)$ in isothermal coordinates.

Astala and Päivärinta [12] reconstructed σ under the assumption $C < \sigma(x) < C^{-1}$ for a constant $C > 0$ without imposing any regularity on σ. They converted the inverse conductivity problem to the following Beltrami equation (2.14). Without loss of generality, we can assume that Ω is a unit disc in $\mathbf{R}^2$. We extend σ to be 1 outside Ω. For a real solution u to the equation $\nabla \cdot \sigma \nabla u = 0$, put $w_1 = -\sigma \partial_2 u$, $w_2 = \sigma \partial_1 u$, where $\partial_j = \partial/\partial x_j$. Then, since $\partial_2 w_1 = \partial_1 w_2$, there exists v such that $\partial_1 v = -\sigma \partial_2 u$, $\partial_2 v = \sigma \partial_1 u$. Letting $f = u + iv$, we obtain

$$\bar{\partial} f = \mu \overline{\partial f}, \quad \mu = \frac{1-\sigma}{1+\sigma}. \tag{2.14}$$

Consider the case $\sigma = 1$. Then, in the polar coordinates

$$\partial_r v = -\partial_\theta u, \quad \partial_\theta v = \partial_r u, \quad \text{on} \quad \partial\Omega.$$

Since $\Delta v = 0$, v is determined by the D-N map Λ_σ with $\sigma = 1$. This is also true for the general case. The D-N map Λ_σ determines μ-Hilbert transform

$$\mathscr{H}_\mu : u\big|_{\partial\Omega} \to v\big|_{\partial\Omega}$$

and vice versa.

The first step is to construct an exponentially growing solution to (2.14). For $k \in \mathbf{C}$, we seek a solution $f_\mu(z, k)$ in the form

$$f_\mu(z, k) = e^{ikz} M_\mu(z, k),$$

$$M_\mu(z, k) = 1 + O(1/z) \quad \text{as} \quad z \to \infty.$$

Let C and B be the Cauchy transform and Beurling transform

$$Cf(z) = -\frac{1}{\pi} \int_{\mathbf{C}} \frac{f(\zeta)}{\zeta - z} dx dy, \quad \zeta = x + iy,$$

$$Bf(z) = -\text{p.v.} \frac{1}{\pi} \int_{\mathbf{C}} \frac{f(\zeta)}{(\zeta - z)^2} dx dy,$$

respectively, and $\bar{B}$ be defined by $\bar{B}f(z) = \overline{Bf(z)}$. Letting $e_k(z) = e^{i(kz+\bar{k}z)}$ and

$$v = \mu(z, k) = e_{-k}(z)\mu(z), \quad \alpha = \alpha(z, k) = -i\bar{k} e_{-k}(z)\mu(z),$$

we define an operator K by

$$Kf = C(1 - v\overline{B})^{-1}(\alpha\overline{f}).$$

Then, the desired solution is defined by

$$M_\mu(z, k) = 1 + (1 - K)^{-1}K\chi_\Omega,$$

where χ_Ω is the characteristic function of Ω. One can show that the D-N map Λ_σ determines $M_\mu(z, k)$ outside Ω. Furthermore, it determines

$$t_\mu(k) = \frac{1}{\pi} \int_\Omega \mu\partial(e_k M_\mu)dxdy,$$

which is an analogue of the scattering amplitude. Next, we solve the $\overline{\partial}$-equation with respect to k:

$$\frac{\partial}{\partial\overline{k}}u_j = -it_\mu(k)\overline{u_j}, \quad j = 1, 2, \tag{2.15}$$

with the asymptotics

$$u_1 = e^{ikz}(1 + O(1/z)), \quad u_2 = e^{ikz}(i + O(1/z)), \quad z \to \infty.$$

Fixing $z_0 \in \mathbf{C}$ such that $|z_0| > 1$, and write

$$\begin{pmatrix} u_1(z, k) \\ u_1(z, k) \end{pmatrix} = T(z_0)(k) \begin{pmatrix} u_1(z_0, k) \\ u_1(z_0, k) \end{pmatrix}.$$

We then see that Λ_σ determines the 2×2 matrix $T_{z_0}(k)$. We have thus constructed f_μ inside Ω. By virtue of (2.14), μ is computed as $\overline{\partial}f_\mu/\partial f_\mu$.

Astala et al. [13] extended this result for the anisotropic case. Namely, any two-dimensional bounded anisotropic conductivity, i.e. $C < (\sigma_{ij}(x)) < C^{-1}$, is uniquely reconstructed from the D-N map up to a diffeomorphism leaving the boundary invariant.

As is seen in the above proof, the inverse boundary value problem has a deep relation with the inverse scattering problem. One passes the problem to the whole space $\mathbf{R}^n$, and from the knowledge of the exterior problem, one derives the desired information of the interior problem.

The two-dimensional inverse boundary value problem for the Schrödinger operator $-\Delta + q(x)$ remained open for a long time, until Bukhgeim [30] used the CGO solutions of the form $u_1(z, k) = e^{z^2 k}(1 + \psi_1)$, $u_2(z, k) = e^{-\overline{z}^2 k}(1 + \psi_2)$.

Then, if $q_{(x)}$ and $q_2(x)$ have the same D-N map, one has

$$\int_\Omega e^{2i\tau x_1 x_2}(q_1 - q_2)(1 + \psi_1 + \psi_2 + \psi_1\psi_2)dx = 0.$$

Applying the stationary phase method, one then obtains $q_1(x) = q_2(x)$. Note that, since z^2k is harmonic, it is a limiting Carleman weight.

Since any harmonic function can be used as a limiting Carleman weight, one has a lot of freedom to construct CGO solutions. In two-dimensions, the partial data problem has more complete solutions. Let Ω be a bounded domain in $\mathbf{R}^2$ with smooth boundary. Take an open set $\Gamma \subset \partial\Omega$ and let $\Gamma_0 = \partial\Omega \setminus \overline{\Gamma}$. Let C_{q_j} be the set of data $\left(u\big|_\Gamma, \frac{\partial u}{\partial \nu}\big|_\Gamma\right)$, where $u \in H^1(\Omega)$, $(-\Delta + q_j)u = 0$ in Ω and $u\big|_{\Gamma_0} = 0$. By virtue of [81], the following theorem holds.

Theorem 2.8 Let $q_j \in C^{2+\alpha}(\overline{\Omega})$ for some $\alpha > 0$. Then, $q_1 = q_2$ provided $C_{q_1} = C_{q_2}$.

Chapter 3
Multi-Dimensional Gel'fand–Levitan Theory

In this chapter, we consider the inverse problem for multi-dimensional Schrödinger operators appearing in quantum scattering theory. After a brief explanation of the scattering phenomena and related spectral properties, we define wave operators and the S-matrix, and then we review the achievements obtained in the study of multi-dimensional Gel'fand–Levitan theory.

The S-matrix was first introduced by Wheeler [196]. Heisenberg [73] placed it on the basis of the theory of elementary particles. As has been stated in Sect. 1.7, the S-matrix is directly related to the physical experiment. Mathematical properties of the S-matrix were made precise mainly for the non-relativistic quantum mechanics, i.e. for Schrödinger operators. Martin [140] proved that the scattering amplitude can be constructed from the differential cross section when the latter is small.[1] It is a general belief that the S-matrix contains all information of the physical system although it is defined only by the observation at infinity. Therefore, the reconstruction of the potential from the S-matrix is the fundamental problem of quantum scattering theory.

From the middle of 1960s to the middle of 1970s, Faddeev brought a progress for the multi-dimensional inverse scattering problem aiming at extending the one-dimensional Gel'fand–Levitan theory. By introducing a new Green's function of the Helmholtz equation, he found an operator theoretical, or algebraic trick behind the Gel'fand–Levitan theory, the reconstruction procedure of the potential and a formulation of the characterization problem.[2] Some parts of this theory are still formal, hence it is worth studying precise and rigorous statements of the inverse scattering procedure. In the 1980s, a new insight was introduced by the $\overline{\partial}$-theory. Khenkin and Novikov [111] formulated the characterization condition by the $\overline{\partial}$-

---

[1]However, in general, the differential cross section does not determine the scattering amplitude uniquely. See [153].

[2]A necessary and sufficient condition guaranteeing for a given function to be the scattering matrix of some potential

© The Author(s), under exclusive licence to Springer Nature Singapore Pte Ltd. 2020   59
H. Isozaki, *Inverse Spectral and Scattering Theory*, SpringerBriefs
in Mathematical Physics 38, https://doi.org/10.1007/978-981-15-8199-1_3

equation. Nachman [148] obtained a representation formula of the potential by the $\bar{\partial}$-method. Compared with the one-dimensional result, the characterization problem in multi-dimension remains as an open big issue.

3.1 Spectra and Scattering Phenomena

In quantum mechanics, any physical phenomenon is described by a self-adjoint operator H in a Hilbert space $\mathscr{H}$. A physical state is represented by a vector $\psi \in \mathscr{H}$, whose behavior in time is governed by the Schrödinger equation $i\frac{\partial}{\partial t}\psi(t) = H\psi(t), \psi(0) = f$. Letting $E(\lambda)$ be the spectral decomposition of H, $\psi(t)$ is written as $\psi(t) = e^{-itH}f = \int_{-\infty}^{\infty} e^{-it\lambda}dE(\lambda)f$. If f is an eigenvector of H with eigenvalue λ, then $\psi(t) = e^{-it\lambda}f$. If $f \in \mathscr{H}_{ac}(H)$, one can show that $\|Ke^{-itH}f\| \to 0$ as $t \to \pm\infty$ for any compact operator K.[3] These simple facts suggest that spectral properties of H give a big influence on the behavior of $\psi(t)$. Let us consider it more precisely for the case $\mathscr{H} = L^2(\mathbf{R}^n)$. For $R > 0$, let $F(|x| < R)$ and $F(|x| > R)$ be the characteristic functions of the set $\{|x| < R\}$ and $\{|x| > R\}$. We say that a self-adjoint operator A in $L^2(\mathbf{R}^n)$ has local-compactness property if $F(|x| < R)(A + i)^{-1}$ is compact for any $R > 0$. The following fact is proved by Ruelle [172] and Amrein and Georgescu [8].

Lemma 3.1 *Assume that a self-adjoint operator A in $L^2(\mathbf{R}^n)$ has local-compactness property. Then, we have*

$$f \in \mathscr{H}_{pp}(A) \Longleftrightarrow \sup_{t>0} \|F(|x| > R)e^{-itA}f\| \to 0 \quad as \quad R \to \infty,$$

$$f \in \mathscr{H}_{cont}(A) \Longleftrightarrow \lim_{T\to\infty} \frac{1}{T} \int_0^T \|F(|x| < R)e^{-itA}f\|^2 dt = 0, \quad \forall R > 0.$$

Therefore, it is natural to call $f \in \mathscr{H}_{pp}(A)$ bound states and $f \in \mathscr{H}_{cont}(A)$ scattering states, since in the latter case, $e^{-itA}f$ decays to 0 on any bounded sets in $\mathbf{R}^n$, hence it is observed only near infinity because of the unitarity of e^{-itA}.

Now let us consider $H = -\Delta + V(x)$, where $V(x)$ is a real function decaying rapidly at infinity. If $f \in \mathscr{H}_{cont}(H)$, by Lemma 3.1, the state $e^{-itH}f$ propagates to infinity. Since $V(x)$ is small near infinity, $e^{-itH}f$ will be governed only by $H_0 = -\Delta$. This means that there will be $f_0^{(+)}$ such that

$$\|e^{-itH}f - e^{-itH_0}f_0^{(+)}\| \to 0, \quad t \to \infty.$$

---

[3]Since $f \in \mathscr{H}_{ac}(H)$, $(e^{-itH}f, g) \to 0$ as $t \to \pm\infty$ for any $g \in \mathscr{H}$.

By the unitarity of e^{-itH}, this implies

$$\| f - e^{itH} e^{-itH_0} f_0^{(+)} \| \to 0, \quad t \to \infty.$$

Considering the case $t \to -\infty$, one also has

$$\| f - e^{itH} e^{-itH_0} f_0^{(-)} \| \to 0, \quad t \to -\infty.$$

This observation raises the following question.

1. The limit $W_{\pm} = \lim_{t \to \pm\infty} e^{itH} e^{-itH_0}$ exists?

If so, any free state $e^{-itH_0} f_0^{(\pm)}$ near infinity is regarded as a limit of a perturbed state $e^{-itH} f$. Then the next question is raised.

2. The ranges of $W_{\pm}$ coincide?

If so, any free state $e^{-itH_0} f_0^{(-)}$ in the remote past evolves again to a free state $e^{-itH_0} f_0^{(+)}$ in the remote future. One can then consider the *scattering operator*

$$S : f_0^{(-)} \to f_0^{(+)},$$

which assigns the outgoing state at infinity in the remote future to the incoming state in the remote past. It turned out that, mathematically, the strong limit is appropriate to deal with the limit for $W_{\pm}$. Moreover, if $W_{\pm}$ exists, one can show that Ran $W_{\pm} \subset \mathscr{H}_{ac}(H)$. Therefore, (2) is guaranteed if we show

$$\text{Ran } W_{\pm} = \mathscr{H}_{ac}(H).$$

This property is called *asymptotic completeness*. Existence and asymptotic completeness of wave operators constitute the main part of the so-called forward problem of scattering theory, and plenty of works have been presented. See, e.g., [167] or [197] for a survey. To study the properties of S, it is convenient to consider its Fourier transform:

$$\widehat{S} = \mathscr{F}_0 S \mathscr{F}_0^*,$$

where $(\mathscr{F}_0 f)(\xi) = (2\pi)^{-n/2} \int_{\mathbf{R}^n} e^{-ix \cdot \xi} f(x) dx$. Taking account of energy conservation, one can prove that for a.e. energy $\lambda = k^2$, there is a unitary operator $\widehat{S}(k)$ on $L^2(S^{n-1})$ such that

$$(\widehat{S}\psi)(k\omega) = (\widehat{S}(k)\psi(k\cdot))(\omega), \quad a.e. \ k > 0, \quad \omega \in S^{n-1},$$

for $\psi(\xi) \in L^2(\mathbf{R}^n)$. The operator $\widehat{S}(k)$ is called Heisenberg's S-matrix.

As we have seen above, scattering is a time-dependent process, and the forward problem can be solved by purely time-dependent methods. However, to study the

detailed structure of the S-matrix and solve the inverse problem, the knowledge of
the resolvent of the Schrödinger operator is indispensable.

3.2 Spectral Theory for Schrödinger Operators

We consider the Schrödinger operator $H = -\Delta + V(x)$ in $\mathbf{R}^n$ ($n \geq 2$), where the
potential $V(x)$ is a real-valued function satisfying

$$|V(x)| \leq C(1 + |x|)^{-\delta_0} \quad \text{for} \quad |x| > R_0 \tag{3.1}$$

for constants $\delta_0 > 1$ and $R_0 > 0$. One can also deal with the case $0 < \delta_0 \leq 1$
assuming differentiability on $V(x)$, which is omitted here for the sake of simplicity.
On the region $|x| \leq R_0$, we can allow mild singularities for $V(x)$ so that

- the essential self-adjointness of $\left(-\Delta + V(x) \right)\big|_{C_0^\infty(\mathbf{R}^n)}$,
- the unique continuation theorem on $\mathbf{R}^n$

are guaranteed. Let $H = -\Delta + V(x)$ and $R(z) = (H - z)^{-1}$ for $z \in \mathbf{C} \setminus \mathbf{R}$. We
then have

$$\sigma_d(H) \subset (-\infty, 0), \quad \sigma_e(H) = [0, \infty).$$

In this case $\sigma_e(H) = \sigma_c(H)$. By the hypothesis of quantum mechanics, H should
possess a complete family of generalized eigenfunctions describing the continuous
spectrum. The first step to construct them is the limiting absorption principle.

For a self-adjoint operator A on a Hilbert space $\mathscr{H}$, the resolvent $(A - \lambda)^{-1}$ does
not exist for $\lambda \in \sigma(A)$. However, when $\lambda \in \sigma_c(A)$, it sometimes happens that the
limit $\lim_{\epsilon \downarrow 0}(A - \lambda \mp i\epsilon)^{-1}$ exists as an operator from X to Y, where X and Y are
appropriate Banach spaces such that $X \subset \mathscr{H} \subset Y$, whose norms satisfy $C^{-1}\|\cdot\|_Y \leq$
$\|\cdot\|_{\mathscr{H}} \leq C\|\cdot\|_X$ for a constant $C > 0$. This fact, called the *limiting absorption
principle*, is a crucial step to study the properties of the continuous spectrum of A.

For the Schrödinger operator $H = -\Delta + V(x)$, the weighted L^2 space defined
in (2.8) is often used as X, Y (see, e.g., [1, 146]). Let $R(z) = (H - z)^{-1}$ for $z \notin \mathbf{R}$.

Theorem 3.1 *Let $s > 1/2$. Then, there exists a norm limit $\lim_{\epsilon \downarrow 0} R(\lambda \pm i\epsilon) =: R(\lambda \pm i0)$ as an operator from $L^{2,s}$ to $L^{2,-s}$ for any $\lambda > 0$. Moreover, $R(\lambda \pm i0)$ is a $\mathbf{B}(L^{2,s}; L^{2,-s})$-valued Hölder continuous function of $\lambda > 0$.*

By this theorem and Stone's formula (Theorem 1.7), H has no singular continuous spectrum.[4] We have already discussed the eigenfunction expansion theory
for the continuous spectrum in the one-dimensional case. The multi-dimensional

---

[4]In disordered systems the singular continuous spectrum appears.

case was first studied by Povzner [160] and Ikebe [78]. It was further extended by Agmon [1], Kato and Kuroda [105, 124], and Saito [175]. As can be checked easily, the Fourier transformation is derived from the asymptotic expansion of the Green function of $-\Delta$ at infinity. The following Besov type space introduced by Agmon and Hörmander [2] is more suitable to describe the behavior of $R(\lambda \pm i0)$ as $|x| \to \infty$.

Letting $R_j = 2^j$ for $j \geq 0$, and $R_{-1} = 0$, we put

$$\Omega_j = \{x \in \mathbf{R}^n : R_{j-1} < |x| < R_j\}, \quad j \geq 0,$$

and let $\mathscr{B}$ be the Banach space equipped with the following norm

$$\|f\|_{\mathscr{B}} = \sum_{j=0}^{\infty} R_j^{1/2} \|f\|_{L^2(\Omega_j)}.$$

The dual space of $\mathscr{B}$ is identified with the set of functions u such that

$$\|u\|_{\mathscr{B}^*} = \sup_{j \geq 0} R_j^{-1/2} \|u\|_{L^2(\Omega_j)} < \infty.$$

There exists a constant $C > 0$ such that

$$C^{-1}\|u\|_{\mathscr{B}^*} \leq \left(\sup_{R > 1} \frac{1}{R} \int_{|x| < R} |u(x)|^2 dx\right)^{1/2} \leq C\|u\|_{\mathscr{B}^*}. \tag{3.2}$$

Therefore from now on, we adopt the middle term of the inequalities of (3.2) as the norm of $\mathscr{B}^*$. The following inclusion relations hold for any $s > 1/2$

$$L^{2,s} \subset \mathscr{B} \subset L^{2,1/2} \subset L^2 \subset L^{2,-1/2} \subset \mathscr{B}^* \subset L^{2,-s}.$$

Define the restriction of Fourier transform by

$$(U_0(k)f)(\omega) = (2\pi)^{-n/2} \int_{\mathbf{R}^n} e^{-ik\omega \cdot x} f(x) dx.$$

Then, for any compact interval $I \subset (0, \infty)$, there exists a constant $C > 0$ such that

$$\|U_0(k)f\|_{L^2(S^{n-1})} \leq C\|f\|_{\mathscr{B}}, \quad k \in I.$$

A remarkable property of the space $\mathscr{B}$ is that $U_0(k) : \mathscr{B} \to L^2(S^{n-1})$ is onto:

$$U_0(k)\mathscr{B} = L^2(S^{n-1}).$$

The adjoint operator $U_0(k)^* : L^2(S^{n-1}) \to \mathscr{B}^*$ characterizes the solution space to the Helmholtz equation:

$$\{u \in \mathscr{B}^* ; \; (-\Delta - k^2)u = 0\} = U_0(k)^* L^2(S^{n-1}).$$

For example, in $\mathbf{R}^3$, letting $R_0(z) = (-\Delta - z)^{-1}$, $R_0(\lambda \pm i0)$ has an integral kernel $\frac{e^{\pm i\sqrt{\lambda}|x-y|}}{4\pi|x-y|}$, which belongs to $\mathscr{B}^*$ for any fixed $y \in \mathbf{R}^3$. It suggests that the spaces $\mathscr{B}$, $\mathscr{B}^*$ are most suitable to deal with the resolvent for the Laplacian. This is true also for $H = -\Delta + V$. The following theorem is due to [2].

Theorem 3.2 *For any $f, g \in \mathscr{B}$ and $\lambda > 0$, there exists a limit*

$$\lim_{\epsilon \downarrow 0}(R(\lambda \pm i\epsilon)f, g) =: (R(\lambda \pm i0)f, g).$$

For any compact interval $I \subset (0, \infty)$, there exists a constant $C > 0$ such that $\|R(\lambda \pm i0)\|_{\mathbf{B}(\mathscr{B}, \mathscr{B}^*)} \leq C$ for $\lambda \in I$.

Based on this theorem, we formulate the eigenfunction expansion theory in the following way. Define

$$\big(\mathscr{F}_0^{(\pm)}(\lambda)f\big)(\omega) = C_{\pm}(\lambda)\big(U_0(\sqrt{\lambda})f\big)(\pm\omega), \quad \omega \in S^{n-1},$$

$$C_{\pm}(\lambda) = e^{\mp(n-3)\pi i/4} 2^{-1/2} \lambda^{(n-2)/4},$$

and put

$$\mathscr{F}^{(\pm)}(\lambda) = \mathscr{F}_0^{(\pm)}(\lambda)\big(1 - VR(\lambda \pm i0)\big).$$

Note

$$\mathscr{F}^{(\pm)}(\lambda) \in \mathbf{B}(\mathscr{B}; L^2(S^{n-1})).$$

For $u, v \in \mathscr{B}^*$, we define

$$u \simeq v \Longleftrightarrow \frac{1}{R}\int_{|x|<R}|u(x) - v(x)|^2 dx \to 0 \quad (R \to \infty).$$

Lemma 3.2 *For any $\lambda > 0$ and $f \in \mathscr{B}$,*

$$R(\lambda \pm i0)f \simeq \Big(\frac{\pi}{\sqrt{\lambda}}\Big)^{1/2}\frac{e^{\pm i\sqrt{\lambda}r}}{r^{(n-1)/2}}\mathscr{F}^{(\pm)}(\lambda)f.$$

In fact, the boundary value of the resolvent $(-\Delta - \lambda - i0)^{-1}$ has an integral kernel $\frac{i}{4}(\frac{\sqrt{\lambda}}{2\pi r})^{(n-2)/2} H^{(1)}_{(n-2)/2}(\sqrt{\lambda} r)$, $r = |x - y|$, where $H^{(1)}_v$ is the Hankel function of the first kind. The asymptotic expansion of the Hankel function then implies the lemma for H_0. The case for H is proven by using the resolvent equation $R(z) = R_0(z) - R_0(z)VR(z)$.

Lemma 3.2 yields $u_\pm = R(\lambda \pm i0)f$ satisfies $(\partial_r - i\sqrt{\lambda})u_\pm \simeq 0$. In this case, $u_\pm$ is said to satisfy the *radiation condition* (outgoing for $+$, incoming for $-$). This is the boundary condition at infinity. The solution of the Schrödinger equation satisfying the radiation condition is unique.

The physical meaning of radiation condition is as follows. For the sake of simplicity, let us consider the case $n = 3$. Then, $u_\pm = e^{\pm ikr}/r$ satisfies the Helmholtz equation $(-\Delta - k^2)u_\pm = 0$ for $x \neq 0$. Hence, $v_\pm = e^{-ikt}u = e^{-ik(t \mp r)}/r$ satisfies the wave equation $(\partial_t^2 - \Delta)v_\pm = 0$. Then, the wave front of v_+, i.e. $\{x ; t - r = C\}$, goes to infinity as $t \to \infty$. By this reason, v_+ is called an outgoing wave. Similarly, v_- is called an incoming wave, since its wave front comes to the origin as $t \to \infty$. Dropping the time-periodic factor e^{-ikt}, $u_\pm$ satisfies $(\partial_r \mp ik)u_\pm = O(r^{-2})$, hence can be neglected. Therefore we say that a function $w(x)$ satisfies the outgoing radiation condition if $(\partial_r - ik)w = o(r^{-1})$, and incoming radiation condition when $(\partial_r + ik)w = o(r^{-1})$.

Letting $u = R(\lambda + i0)f$, $v = R(\lambda + i0)g$ and applying Green's formula for the integral $\int_{|x|<R} ((\Delta u)\overline{v} - u\Delta\overline{v})\, dx$, we obtain the following lemma.

Lemma 3.3 *For any $f, g \in \mathscr{B}$,*

$$\frac{1}{2\pi i}\big((R(\lambda + i0) - R(\lambda - i0))f, g\big) = \big(\mathscr{F}^{(\pm)}(\lambda)f, \mathscr{F}^{(\pm)}(\lambda)g\big)_{L^2(S^{n-1})}.$$

For $f \in \mathscr{B}$, put

$$(\mathscr{F}^{(\pm)}f)(\lambda) = \mathscr{F}^{(\pm)}(\lambda)f.$$

Let $\mathbf{h} = L^2(S^{n-1})$, and $\mathbf{H} = L^2((0, \infty); \mathbf{h})$ be the set of $\mathbf{h}$-valued L^2-functions on $(0, \infty)$ with respect to $d\lambda$. Let $\mathscr{H} = L^2(\mathbf{R}^n)$.

Theorem 3.3

(1) $\mathscr{F}^{(\pm)}$ is uniquely extended to a partial isometry with initial set $\mathscr{H}_{ac}(H)$ and final set $\mathbf{H}$.[5]

(2) For $f \in D(H)$, $(\mathscr{F}^{(\pm)}Hf)(\lambda) = \lambda(\mathscr{F}^{(\pm)}f)(\lambda)$, a.e. $\lambda > 0$.

---

[5]Given two Hilbert spaces $\mathscr{H}_i$ and closed subspaces $S_i \subset \mathscr{H}_i$, $i = 1, 2$, a linear operator $A : \mathscr{H}_1 \to \mathscr{H}_2$ is said to be a partial isometry with initial set S_1 and final set S_2 if $Ax = 0$ for $x \in S_1^\perp$ and $\|Ax\| = \|x\|$ for $x \in S_1$, moreover $AS_1 = S_2$.

(3) $\mathscr{F}^{(\pm)}(\lambda)^* \in \mathbf{B}(\mathbf{h}; \mathscr{B}^*)$ is an eigen operator of H in the sense that

$$(-\Delta + V(x) - \lambda)\mathscr{F}^{(\pm)}(\lambda)^*\phi = 0$$

holds for any $\phi \in \mathbf{h}$.
(4) For any $f \in \mathscr{H}_{ac}(H)$ and $0 < a < b < \infty$, $\int_a^b \mathscr{F}^{(\pm)}(\lambda)^*\big(\mathscr{F}^{(\pm)}f\big)(\lambda)d\lambda$, defined as $\mathscr{B}^*$-valued integral, belongs to $\mathscr{H}$ and

$$\lim_{N \to \infty} \int_{1/N}^N \mathscr{F}^{(\pm)}(\lambda)^*\mathscr{F}^{(\pm)}(\lambda)\hat{f}d\lambda = f$$

holds in the norm of $\mathscr{H}$.

The wave operator $W_\pm = \mathrm{s} - \lim_{t \to \pm\infty} e^{itH}e^{-itH_0}$ exists and is asymptotically complete. In fact, letting $\mathscr{F}_0 = \mathscr{F}_0^{(+)}$, we have $W_\pm = \big(\mathscr{F}^{(\pm)}\big)^*\mathscr{F}_0$. Then, $\mathrm{Ran}\, W_\pm = \mathscr{H}_{ac}(H)$ by Theorem 3.3 (1). The scattering operator and its Fourier transform are defined by

$$S = \big(W_+\big)^*W_-, \quad \widehat{S} = \mathscr{F}_0 S \big(\mathscr{F}_0\big)^*.$$

It then has the following representation

$$\big(\widehat{S}\widehat{f}\big)(\lambda) = \widehat{S}(\lambda)\widehat{f}(\lambda), \quad \lambda > 0,$$

for any $\widehat{f} \in \mathbf{H}$, where $\widehat{S}(\lambda)$ is a unitary operator on $L^2(S^{n-1})$ and written as

$$\widehat{S}(\lambda) = 1 - 2\pi i A(\lambda), \tag{3.3}$$

$$A(\lambda) = \mathscr{F}_0(\lambda)V\mathscr{F}_0(\lambda)^* - \mathscr{F}_0(\lambda)VR(\lambda + i0)V\mathscr{F}_0(\lambda)^*. \tag{3.4}$$

For any $\lambda > 0$, we put

$$\mathscr{N}(\lambda) = \{u \in \mathscr{B}^*\,;\, (-\Delta + V - \lambda)u = 0\}.$$

Then, we have

$$\mathscr{N}(\lambda) = \mathscr{F}^{(\pm)}(\lambda)^*\mathbf{h}.$$

We put $(J\phi)(\omega) = \phi(-\omega)$, $\phi \in \mathbf{h}$, and define the geometric S-matrix $\widehat{S}_{geo}(\lambda)$ by

$$\widehat{S}_{geo}(\lambda) = \widehat{S}(\lambda)J.$$

Theorem 3.4 *For any $\varphi^{(in)} \in \mathbf{h}$, there exist unique $u \in \mathcal{N}(\lambda)$ and $\varphi^{(out)} \in \mathbf{h}$ such that*

$$u \simeq C_+(\lambda)\frac{e^{i\sqrt{\lambda}r}}{r^{(n-1)/2}}\varphi^{(out)} - C_-(\lambda)\frac{e^{-i\sqrt{\lambda}r}}{r^{(n-1)/2}}\varphi^{(in)}.$$

Moreover

$$\varphi^{(out)} = \widehat{S}_{geo}(\lambda)\varphi^{(in)}.$$

For the proof of the above facts, see [76, 197] and [91].

3.3 Generalized Eigenfunctions

Let us derive the classical eigenfunction expansion theorem in the Sect. 1.7 from the results in the previous subsection. Assume that $V(x) \in L^{2,s}$ for some $s > 1/2$, i.e. $\delta_0 > (n+1)/2$, where δ_0 is in (3.1). We put

$$\varphi_\pm(x,\xi) = \varphi_0(x,\xi) - \big(R(|\xi|^2 \pm i0)(V(\cdot)\varphi_0(\cdot,\xi))(x), \quad \varphi_0(x,\xi) = e^{ix\cdot\xi}.$$

It satisfies the Schrödinger equation

$$\big(-\Delta + V(x) - |\xi|^2\big)\varphi_\pm(x,\xi) = 0.$$

A simple manipulation shows that

$$\big(\mathscr{F}^{(\pm)}(\lambda)f\big)(\omega) = (2\pi)^{-n/2}C_\pm(\lambda)\int_{\mathbf{R}^n}\overline{\varphi_\mp(x,\mp\sqrt{\lambda}\omega)}f(x)dx.$$

Using

$$A(\lambda) = \mathscr{F}_0(\lambda)V\big(1 - R(\lambda+i0)V\big)\mathscr{F}_0(\lambda)^* = \mathscr{F}^{(+)}(\lambda)V\mathscr{F}_0(\lambda)^*,$$

we see that $A(\lambda)$ has an integral kernel

$$A(\lambda,\theta,\omega) = \frac{(\sqrt{\lambda})^{n-2}}{2}(2\pi)^{-n}\int_{\mathbf{R}^n}\varphi_+(x,\sqrt{\lambda}\theta)V(x)e^{i\sqrt{\lambda}\omega\cdot x}dx.$$

Lemma 3.2 implies

$$\varphi_+(x,\sqrt{\lambda}\omega) - e^{i\sqrt{\lambda}\omega\cdot x} \sim -C_1(\lambda)\frac{e^{i\sqrt{\lambda}r}}{r^{(n-1)/2}}a(\lambda,\theta,\omega),$$

where $r = |x|, \theta = x/r$ and

$$C_1(\lambda) = e^{-(n-3)\pi i/4} 2^{-(n+1)/2} \pi^{-(n-1)/2} \lambda^{(n-3)/4},$$

$$a(\lambda, \theta, \omega) = \int_{\mathbf{R}^n} \varphi_+(x, \sqrt{\lambda}\theta) V(x) e^{i\sqrt{\lambda}\omega \cdot x} dx.$$

Therefore, $a(\lambda, \theta, \omega)$ and $A(\lambda, \theta, \omega)$ coincide up to a constant depending only on λ.

When $V(x)$ is compactly supported, the scattering amplitude has an analytic continuation with respect to $k = \sqrt{\lambda}$ onto the upper half-plane $\{\Im k > 0\}$ with poles on the imaginary axis $k_1 = i\gamma_1, k_2 = i\gamma_2, \cdots$. Moreover, $k_j^2 = -\gamma_j^2$ are the negative eigenvalues of $-\Delta + V(x)$.

3.4 The Role of Volterra Integral Operator

Let us repeat the essential part of the one-dimensional Gel'fand–Levitan theory. Let $\varphi(x, k)$ be the solution of the problem

$$-\varphi''(x, k) + V(x)\varphi(x, k) = k^2 \varphi(x, k), \quad x > 0, \tag{3.5}$$

$$\varphi(0, k) = 0, \quad \varphi'(0, k) = 1.$$

Then, $\varphi(x, k)$ is an even and entire function of $k \in \mathbf{C}$ satisfying

$$\varphi(x, k) = \frac{\sin kx}{k} + o\left(\frac{e^{|\mathrm{Im}\,k|x}}{|k|}\right), \quad |k| \to \infty.$$

By the Paley–Wiener theorem, $\varphi(x, k)$ has the following representation

$$\varphi(x, k) = \frac{\sin kx}{k} + \int_0^x K(x, y) \frac{\sin ky}{k} dy. \tag{3.6}$$

Putting it into (3.5), we obtain the equation

$$\left(\partial_y^2 - \partial_x^2 + V(x)\right) K(x, y) = 0. \tag{3.7}$$

The crucial fact is that

$$K(x, y) + \Omega(x, y) + \int_0^x K(x, t)\Omega(t, y) dt = 0, \quad x > y, \tag{3.8}$$

$$2 \frac{d}{dx}\left(K(x, x)\right) = V(x), \tag{3.9}$$

where $\Omega(x, y)$ is a function constructed from the S-matrix and information of bound states. By solving the Eq. (3.8), we obtain $K(x, y)$, from which we can compute $V(x)$ by using (3.9). Therefore, the key formula is (3.8).

What is the hidden mechanism of this argument? Kay and Moses [108] studied the algebraic aspect of Gel'fand–Levitan theory. Their crucial idea is to utilize the intertwining property. Let A, B be self-adjoint operators. A bounded operator T from $\mathcal{H}_{ac}(A)$ to $\mathcal{H}_{ac}(B)$ is said to *intertwine* A and B if it satisfies $AT = TB$. A well-known example is the spectral representation. Assume that $\sigma_e(A) = \sigma_e(B) =: I$. A unitary operator F from $\mathcal{H}_{ac}(A)$ to $L^2(I; \mathbf{h})$, $\mathbf{h}$ being an auxiliary Hilbert space, is called a spectral representation of A if

$$(FAu)(\lambda) = \lambda(Fu)(\lambda), \quad \lambda \in I, \quad u \in \mathcal{H}_{ac}(A).$$

We have constructed spectral representations $\mathcal{F}_0$ and $\mathcal{F}$ for Schrödinger operators $H_0 = -d^2/dx^2$ and $H = -d^2/dx^2 + V(x)$. It is obvious that $W = \mathcal{F}^* \mathcal{F}_0$ intertwines H_0 and H. We then have by (1.51)

$$W(x, y) = \frac{2}{\pi} \int_0^\infty k\varphi(x, k) \sin ky \frac{dk}{|F(k)|^2},$$

where $F(k)$ is defined by (1.55). The intertwining property implies that the integral kernel $W(x, y)$ of W satisfies

$$\left(-\partial_x^2 + V(x) \right) W(x, y) = -\partial_y^2 W(x, y).$$

As we have seen in Chap. 1, the essential tool for the one-dimensional inverse problem is the transformation operator. Although the arguments below are formal in some parts, they explain the basic idea behind the Gel'fand–Levitan theory.

Let us consider the operator T with the following integral kernel

$$T(x, y) = \frac{2}{\pi} \int_0^\infty k\varphi(x, k) \sin ky\, dk. \tag{3.10}$$

Then, T intertwines H_0 and H, hence $T(x, y)$ satisfies the equation

$$\left(-\partial_x^2 + V(x) \right) T(x, y) = -\partial_y^2 T(x, y).$$

By virtue of (3.6), we have

$$T(x, y) = \delta(x - y) + \eta(x - y) K(x, y),$$

where $\eta(t)$ is the Heaviside function. Namely, T is a Volterra integral operator. Taking account of Lemma 1.8, T is the transformation operator for H_0 and H. The key fact discovered by Kay and Moses [108] is the following theorem.

Theorem 3.5 *Let $H_0 = -d^2/dx^2$ be defined on $(0, \infty)$ with Dirichlet boundary condition, and Q a self-adjoint operator. Put $H = H_0 + Q$. Let $U = I + K$ be a bounded operator on $L^2(0, \infty)$, where K has a real triangular kernel:*

$$(Ku)(x) = \int_0^x K(x, y)u(y)dy,$$

and $K(x, 0) = 0$. Assume that

$$HU = UH_0 \quad on \quad H^2(0, \infty) \cap H_0^1(0, \infty). \tag{3.11}$$

Then, Q is an operator of multiplication by

$$q(x) = 2\frac{d}{dx}\Big(K(x, x)\Big).$$

For $x > y$, the following equation holds:

$$(\partial_x^2 - \partial_y^2)K(x, y) = q(x)K(x, y). \tag{3.12}$$

Proof By virtue of (3.11), we have

$$QU = D + A, \tag{3.13}$$

where D is an operator of multiplication by $2\frac{d}{dx}K(x, x)$ and A is an integral operator with triangular kernel $K_{xx}(x, y) - K_{yy}(x, y)$. Since U is a Volterra operator, so is its inverse $U^{-1} = 1 + L$, where L has a triangular kernel. Then, we have

$$Q - D = DL + A + AL.$$

The right-hand side has a triangular kernel, while the left-hand side is self-adjoint. Then, we have $Q - D = 0$. The Eq. (3.12) follows from (3.13). As is easily seen, this theorem is also true for complex-valued $K(x, y)$.[6]

It would thus be useful to generalize the notion of Volterra operator to higher dimensions. Let $\gamma \in S^{n-1}$ be arbitrarily fixed. An integral operator K is said to be triangular with respect to γ if its kernel $K(x, y)$ satisfies

$$K(x, y) = 0 \quad for \quad x \cdot \gamma < y \cdot \gamma.$$

---

[6]This proof is formal in the sense that precise conditions on the kernel $K(x, y)$ are lacking. We do not pursue them, however, since we are now interested in the algebraic aspect of Gel'fand–Levitan theory.

We call $U = I + K$ a Volterra operator with respect to γ if K is triangular with respect to γ. The following theorem is proven in the same way as in Theorem 3.5. We denote $x \in \mathbf{R}^n$ as $x = (x_1, x')$.

Theorem 3.6 *Let $H_0 = -\Delta$ on $\mathbf{R}^n$, and $H = H_0 + Q$, where Q is self-adjoint. Suppose $U = I + K$ intertwines H_0 and H. If K is triangular with respect to $(1, 0, \cdots, 0)$, Q is an integral operator with respect to the variable x' and a multiplication operator with respect to x_1, i.e.*

$$(Qu)(x) = \int_{\mathbf{R}^{d-1}} Q(x_1, x', y')u(x_1, y')dy',$$

and its kernel is given by

$$Q(x_1, x', y') = 2\frac{d}{dx_1}\Big(K(x_1, x', x_1, y')\Big).$$

Moreover, for $x_1 > y_1$

$$(\Delta_x - \Delta_y)K(x, y) = \int_{\mathbf{R}^{n-1}} Q(x_1, x', z')K(x_1, z', y_1, y')dz'.$$

We have now arrived at a crucial point of Gel'fand–Levitan theory. In the one-dimensional case (the half-line case), we constructed the generalized eigenfunction of $-d^2/dx^2 + V(x)$ by solving the initial value problem. This solution behaves like a sum of outgoing and incoming waves at infinity, hence corresponds to the physical phenomenon. Then, the intertwining operator T with kernel (3.10) becomes the Volterra type and the potential $V(x)$ is reconstructed from its kernel. The situation is not the same for the multi-dimensional problem. One can construct the physical eigenfunction, however, it is not obvious to construct a Volterra intertwining operator. The reason is that in the one-dimensional case, one is sitting on the real line, which has an obvious direction, while in the multi-dimensional case, there is no peculiar direction.

The remedy consists in abandoning physical eigenfunctions and in looking for non-physical eigenfunctions by which the intertwining operator becomes Volterra type. Kay and Moses tried to find the Volterra operator with respect to the radial variable, however, not successful. It was Faddeev who proceeded to an essential step in realizing this idea [59–61].

3.5 Faddeev Theory

Let us follow the outline of Faddeev theory. Letting $\widetilde{G}(\zeta)$ be the Faddeev Green operator in (2.7), we put $u = (1 + \widetilde{G}(\zeta)V)^{-1}\widetilde{G}(\zeta)f$, which is a solution to $(-\Delta + V - E)u = f$. Any $\zeta \in \mathbf{C}^n$ with $\text{Im}\,\zeta \neq 0$ can be uniquely written as $\zeta = \eta + z\gamma$, where $\eta \in \mathbf{R}^n$, $\gamma \in S^{n-1}$, $\eta \cdot \gamma = 0$, and $z \in \mathbf{C}_+ = \{\operatorname{Im} z > 0\}$. When $z \to t \in \mathbf{R}$, $\widetilde{G}(\zeta)$ tends to $\widetilde{G}(k)$, where $k = \eta + t\gamma \in \mathbf{R}^n$ with $k^2 = E$. We have thus obtained the following solution

$$\psi = e^{ik \cdot x} - e^{ik \cdot x}(1 + \widetilde{G}(k)V)^{-1}\widetilde{G}(k)V \tag{3.14}$$

of $(-\Delta + V - E)\psi = 0$. An important feature of Faddeev's Green operator is that $\widetilde{G}(\eta + z\gamma)$ is analytic with respect to $z \in \mathbf{C}_+$. Then, the 2nd term of the right-hand side of (3.14) is a boundary value of a function analytic in the upper half-plane. The Paley–Wiener theorem then implies that ψ has the following expression

$$\psi = e^{ik \cdot x} - \int_{\gamma \cdot x}^{\infty} A_\gamma(x, \eta, s)e^{ist}ds.$$

This is the Volterra integral equation we were seeking.

With this eigenfunction ψ, one can construct a spectral representation of H and the associated intertwining operator of H_0 and H. Since ψ has triangular expression, this intertwining operator will be of Volterra type. One can then expect an analogue of the one-dimensional Gel'fand–Levitan theory.

Faddeev proceeded as follows. The basic strategy is to replace the usual Green operator by the direction dependent Green operator (to be defined later), which we denote by $R_\gamma(E, t)$. Using the non-physical eigenfunction

$$\Psi_\gamma(x, E, \theta) = e^{i\sqrt{E}\theta \cdot x} - R_\gamma(E, \sqrt{E}\theta \cdot \gamma)(Ve^{i\sqrt{E}\theta \cdot}),$$

we define a new scattering amplitude

$$\widetilde{A}_\gamma(E, \theta, \theta') = (2\pi)^{-n}2^{-1}E^{(n-2)/2}\int_{\mathbf{R}^n} e^{-i\sqrt{E}\theta \cdot x}V(x)\Psi_\gamma(x, E, \theta')dx.$$

We put

$$Q_\gamma^{(\pm)}(E, \theta, \theta') = 2\pi i \chi(\pm\gamma \cdot (\theta - \theta') \geq 0)\widetilde{A}_\gamma(E, \theta, \theta'),$$

where $\chi(\cdots)$ denotes the characteristic function of the set $\{\cdots\}$. Letting $Q_\gamma^{(\pm)}(E)$ be the integral operator with kernel $Q_\gamma^{(\pm)}(E, \theta, \theta')$, Faddeev derived the following conclusions.

1. *Factorization of the S-matrix*

$$\widehat{S}(E) = (1 - Q_\gamma^{(-)}(E))(1 + Q_\gamma^{(+)}(E))^{-1}. \tag{3.15}$$

2. *Volterra operator.* Let $\Phi_\gamma(x, \xi) = \Psi_\gamma(x, |\xi|^2, \xi/|\xi|)$ and put

$$U_\gamma(x, y) = (2\pi)^{-n}\int_{\mathbf{R}^n} \Phi_\gamma(x, \xi)e^{-iy \cdot \xi}d\xi.$$

Then, we have

$$U_\gamma(x, y) = \delta(x - y) - K_\gamma(x, y),$$
$$K_\gamma(x, y) = 0 \quad \text{if} \quad x \cdot \gamma > y \cdot \gamma.$$
(3.16)

3. *Gel'fand–Levitan equation.* For $x \cdot \gamma < y \cdot \gamma$, we have

$$K_\gamma(x, y) + \Omega_\gamma(x, y) + \int_{(x-y)\cdot\gamma < 0} K_\gamma(x, z)\Omega_\gamma(z, y)dz,$$

where $\Omega_\gamma(x, y)$ is a function constructed from the scattering data.

The procedure of the reconstruction of $V(x)$ is as follows. From the scattering matrix $\widehat{S}(E)$, construct $Q_\gamma^{(\pm)}(E)$, and then $\Omega_\gamma(x, y)$. Solve the Gel'fand–Levitan equation to obtain $K_\gamma(x, y)$. Since $I - K_\gamma$ is a Volterra operator, the potential $V(x)$ is reconstructed from $K_\gamma(x, y)$. Faddeev proceeded further. He observed that the analyticity of $\widetilde{G}(\eta + z\gamma)$ plays a key role to guarantee that when we reconstruct $V(x)$ from $\widetilde{A}_\gamma$, $V(x)$ is independent of the artificially introduced direction γ. This will be crucial in the characterization of the scattering matrix.

3.6 Changing Green Operators

In the following sections, we examine Faddeev's theory in detail. It is better to come back to the general setting for the limiting absorption. Let $\mathscr{H}$ be a Hilbert space, and H_0 a self-adjoint operator in $\mathscr{H}$. Assume that there exist two Banach spaces $\mathscr{H}_\pm$ such that $\mathscr{H}_+ \subset \mathscr{H} \subset \mathscr{H}_-$ with dense and continuous inclusions. Assume that there exists an interval $I \subset \sigma_c(H_0)$ such that for any $\lambda \in I$, the following strong limit

$$R_0(\lambda \pm i0) = (H_0 - \lambda \mp i0)^{-1} \in \mathbf{B}(\mathscr{H}_+; \mathscr{H}_-)$$

exists. Suppose there exists an auxiliary Hilbert space $\mathbf{h}$ such that for any $\lambda \in I$, there exists a bounded operator $\mathscr{F}_0(\lambda) \in \mathbf{B}(\mathscr{H}_+; \mathbf{h})$ satisfying

$$\mathscr{F}_0(\lambda)H_0u = \lambda\mathscr{F}_0(\lambda)u, \quad u \in D(H_0) \cap \mathscr{H}_+.$$

We finally assume that the operator $\mathscr{F}_0$ defined by $(\mathscr{F}_0u)(\lambda) = \mathscr{F}_0(\lambda)u$ is extended uniquely to a unitary operator from $\mathscr{H}_0$ to $L^2(I; \mathbf{h})$. Let V be a bounded self-adjoint operator on $\mathscr{H}$ such that $V \in \mathbf{B}(\mathscr{H}_+; \mathscr{H}_+)$, and let $H = H_0 + V$.[7]

---

[7]For the sake of simplicity, we consider here only a perturbation by bounded operators. The extension to relatively bounded perturbations is not difficult.

Definition 3.1 Let $E \in I$. An operator $G^{(0)} \in \mathbf{B}(\mathscr{H}_+; \mathscr{H}_-)$ is said to be a Green operator of $H_0 - E$ if

$$(H_0 - E)G^{(0)} = I \quad \text{on} \quad \mathscr{H}_+. \tag{3.17}$$

An operator $G \in \mathbf{B}(\mathscr{H}_+; \mathscr{H}_-)$ is called a perturbed Green operator associated with $G^{(0)}$ if it satisfies

$$G = G^{(0)} - G^{(0)}VG = G^{(0)} - GVG^{(0)}. \tag{3.18}$$

Note that $(H - E)G = I$ on $\mathscr{H}_+$ by virtue of (3.17) and (3.18). Using a perturbed Green operator G, we define the scattering amplitude associated with G by

$$A = \mathscr{F}_0(E)(V - VGV)\mathscr{F}_0(E)^*.$$

Assume that we are given two Green operators $G_1^{(0)}$ and $G_2^{(0)}$ for $H_0 - E$. Let G_1 and G_2 be the associated perturbed Green operators, and A_1, A_2 the scattering amplitude associated with G_1, G_2, respectively. We then have the following relation between them.

Theorem 3.7 *Assume that there exists $M \in \mathbf{B}(\mathbf{h}; \mathbf{h})$ such that*

$$G_2^{(0)} - G_1^{(0)} = \mathscr{F}_0(E)^* M \mathscr{F}_0(E). \tag{3.19}$$

Then we have

$$A_2 = A_1 - A_1 M A_2. \tag{3.20}$$

In fact, by (3.18), we have

$$(1 - G_1 V)(G_2^{(0)} - G_1^{(0)})(1 - VG_2) = (1 - G_1 V)G_2 - G_1(1 - VG_2) = G_2 - G_1.$$

Multiplying by $\mathscr{F}_0(E)V$ from the left, and $V\mathscr{F}_0(E)^*$ from the right, we obtain the theorem.

Let us discuss the solvability of the Eq. (3.20). Assume that

$$G_1^{(0)}V, \ G_2^{(0)}V : \mathscr{H}_- \to \mathscr{H}_- \text{ and } A_1 : \mathbf{h} \to \mathbf{h} \text{ are compact.}$$

Theorem 3.8 *Under the above assumptions, the Eq. (3.20) is uniquely solvable with respect to A_2:*

$$A_2 = (1 + A_1 M)^{-1} A_1.$$

The proof is again based on simple algebraic manipulations. Since $(1 + G_1^{(0)}V)(1 - G_1 V) = 1$, letting $\widetilde{K} = (1 - G_1 V)(G_2^{(0)} - G_1^{(0)})V$, we have

$$1 + G_2^{(0)} V = (1 + G_1^{(0)} V)(1 + \widetilde{K}). \tag{3.21}$$

The existence of the perturbed Green operator G_j is equivalent to $-1 \notin \sigma_p(G_j^{(0)}V)$. Letting

$$S_1 = (1 - G_1 V)\mathscr{F}_0(E)^* M, \quad S_2 = \mathscr{F}_0(E)V,$$

we have by (3.19)

$$\widetilde{K} = S_1 S_2, \quad A_1 M = S_2 S_1.$$

Since $\sigma_p(S_1 S_2) \setminus \{0\} = \sigma_p(S_2 S_1) \setminus \{0\}$, we have

$$-1 \notin \sigma_p(\widetilde{K}) \iff -1 \notin \sigma_p(A_1 M). \tag{3.22}$$

Theorem 3.8 now follows from (3.20)–(3.22).

We have seen that for a certain pair of Green operators there is a linear equation between the corresponding scattering amplitudes, which is solvable. Now the question is : What kind of property of the Green operator is useful in inverse scattering? According to Faddeev, it is analyticity. We observe it in the next section.

3.7   Direction Dependent Green Operators

There is a delicate issue for the analyticity of Faddeev's Green function. To make it clear, it is convenient to consider the following operator

$$G_{\gamma,0}(\lambda, z)f(x) = (2\pi)^{-n/2} \int_{\mathbf{R}^n} \frac{e^{ix\cdot\xi}}{\xi^2 + 2z\gamma\cdot\xi - \lambda^2} \widehat{f}(\xi)d\xi,$$

where $\gamma \in S^{n-1}$, $\lambda > 0$ and $z \in \mathbf{C}_+$. If $\zeta = \eta + z\gamma$, where $\eta \in \mathbf{R}^n$ and $\eta \cdot \gamma = 0$, we have

$$e^{-ix\cdot\eta}G_{\gamma,0}(|\eta|^2, z)e^{ix\cdot\eta} = \widetilde{G}(\zeta).$$

Theorem 3.9 *Let* $s > 1/2$.

(1) As a $\mathbf{B}(L^{2,s}; L^{2,-s})$*-valued function,* $G_{\gamma,0}(\lambda, z)$ *is continuous with respect to* $\lambda \geq 0$, $\gamma \in S^{n-1}$, $z \in \overline{\mathbf{C}_+}$ *except for* $(\lambda, z) = (0, 0)$.
(2) $G_{\gamma,0}(\lambda, z)$ *is a* $\mathbf{B}(L^{2,s}; L^{2,-s})$*-valued analytic function of* $z \in \mathbf{C}_+$.

(3) For any $\epsilon_0 > 0$, there exists $C > 0$ such that for $\lambda + |z| > \epsilon_0$,

$$\|G_{\gamma,0}(\lambda, z)\|_{\mathbf{B}(L^{2,s};L^{2,-s})} \leq C(\lambda + |z|)^{-1}.$$

(4) For $t \in \mathbf{R}$, let

$$\widetilde{R}_{\gamma,0}(\lambda, t) = e^{it\gamma \cdot x} G_{\gamma,0}(\lambda, t) e^{-it\gamma \cdot x}.$$

Then, we have

$$(-\Delta - \lambda^2 - t^2)\widetilde{R}_{\gamma,0}(\lambda, t) = I.$$

See, e.g., [194] for the proof. Note that the analyticity of $G_{\gamma,0}(\lambda, z)$ is a consequence of the following $\overline{\partial}_z$-formula.

Lemma 3.4 *Let D be an open set in $\mathbf{C}$. Let $p(\xi, z)$ be a $\mathbf{C}$-valued function smooth in $\xi \in \mathbf{R}^n$ and analytic in $z \in D$. Let $M_z = \{\xi \in \mathbf{R}^n ; p(\xi, z) = 0\}$ and assume that for $z \in D$, $\nabla_\xi \mathrm{Re}\, p(\xi, z)$ and $\nabla_\xi \mathrm{Im}\, p(\xi, z)$ are linearly independent on M_z. Then the distribution $S(z)$ defined by*

$$S(z)f = \int_{\mathbf{R}^n} \frac{f(\xi)}{p(\xi, z)} d\xi, \quad f(\xi) \in C_0^\infty(\mathbf{R}^n)$$

satisfies

$$\overline{\partial}_z S(z) = \pi \int_{\mathbf{R}^n} f(\xi)\overline{\partial_z p(\xi, z)}\delta(p(\xi, z))d\xi,$$

where $\delta(p(\xi, z))d\xi$ is the distribution defined by

$$\int_{\mathbf{R}^n} \varphi(\xi)\delta(p(\xi, z))d\xi = \int_{M_z} \varphi(\xi)dM_z, \quad \varphi(\xi) \in C(\mathbf{R}^n),$$

and dM_z is the induced measure on M_z.

Let us give a sketch of the proof of this Lemma. For $\epsilon > 0$, consider

$$S_\epsilon(z)f = \int_{\mathbf{R}^n} \frac{\overline{p(\xi, z)}f(\xi)}{|p(\xi, z)|^2 + \epsilon} d\xi.$$

Then, we have

$$\overline{\partial}_z S_\epsilon(z)f = \int_{\mathbf{R}^n} \frac{\epsilon \overline{\partial_z p}\, f}{(|p|^2 + \epsilon)^2} d\xi.$$

Let $t = \operatorname{Re} p$, $s = \operatorname{Im} p$ and make the change of variable $\xi \to (t, s, \eta)$. Further, put $r = \sqrt{t^2 + s^2}$. Then, we have

$$\overline{\partial_z} S_\epsilon(z) f = \int_{\mathbf{R}^n} \frac{\epsilon \overline{\partial_z p} f}{(r^2 + \epsilon)^2} J r \, dr \, d\theta \, d\eta,$$

J being the Jacobian. Putting $r = \sqrt{\epsilon} \rho$, we then have

$$\overline{\partial_z} S_\epsilon(z) f = \int_{\mathbf{R}^n} \frac{\overline{\partial_z p} f}{(\rho^2 + 1)^2} J \rho \, d\rho \, d\theta \, d\eta.$$

Letting $\epsilon \to 0$, we obtain the lemma.

Another important feature of Faddeev's Green function is the following formula

$$\widetilde{R}_{\gamma,0}(\lambda, t) = R_0(E - i0) M_\gamma^{(+)}(t) + R_0(E + i0) M_\gamma^{(-)}(t), \tag{3.23}$$

where $E = \lambda^2 + t^2$, $R_0(E \pm i0) = (-\Delta - E \mp i0)^{-1}$ and

$$M_\gamma^{(\pm)}(t) = (\mathscr{F}_{x \to \xi})^{-1} \chi(\pm \gamma \cdot (\xi - t\gamma) \geq 0) \mathscr{F}_{x \to \xi},$$

where $\mathscr{F}_{x \to \xi}$ is the Fourier transformation : $f(x) \to \widehat{f}(\xi)$, and $\chi(\pm \gamma \cdot (\xi - t\gamma) \geq 0)$ denotes the operator of multiplication by the characteristic function of the set $\{\xi \in \mathbf{R}^n ; \pm \gamma \cdot (\xi - t\gamma) \geq 0\}$. The formula (3.23) is intuitively obvious, since it has the following (formal) integral kernel

$$\widetilde{R}_{\gamma,0}(\lambda, t) f(x) = (2\pi)^{-n/2} \int_{\mathbf{R}^n} \frac{e^{ix \cdot \xi}}{\xi^2 + 2i0\gamma \cdot (\xi - t\gamma) - E} \widehat{f}(\xi) d\xi.$$

Now, we arrive at a delicate property of Faddeev's Green operator. Let $E = \lambda^2 + t^2$. Then, the integral kernel of $G_{\gamma,0}(\lambda, t)$ is written as

$$(2\pi)^{-n} \int \frac{e^{i(x-y) \cdot \xi}}{\xi^2 + 2(t + i0)\gamma \cdot \xi + t^2 - E} d\xi.$$

One might think that the analytic continuation of this kernel is

$$(2\pi)^{-n} \int \frac{e^{i(x-y) \cdot \xi}}{\xi^2 + 2z\gamma \cdot \xi + z^2 - E} d\xi.$$

However, this is not analytic, which can be seen by Lemma 3.4. The proper analytic continuation of $G_{\gamma,0}(\lambda, t)$ was found by Eskin and Ralston [57]. It consists of two operators:

$$U_{\gamma,0}(E, z) = V_{\gamma,0}(E, z) + W_{\gamma,0}(E, z).$$

Let

$$D_\epsilon = \{z \in \mathbf{C}_+ \,;\; |\mathrm{Re}\, z| < \epsilon/2\}.$$

Take $\varphi_1(t) \in C^\infty(\mathbf{R})$ such that $\varphi_1(t) = 1$ for $|t| > 2\epsilon$, $\varphi_1(t) = 0$ for $|t| < \epsilon$, and put $\varphi_0(t) = 1 - \varphi_1(t)$. For $z \in D_\epsilon$, $V_{\gamma,0}(E, z)$ is defined by

$$V_{\gamma,0}(E, z)f = (2\pi)^{-n/2} \int_{\mathbf{R}^n} \frac{e^{i(x-y)\cdot\xi}\varphi_0(\gamma \cdot \xi)}{\xi^2 + 2z\gamma \cdot \xi + z^2 - E} \widehat{f}(\xi)d\xi.$$

This is $\mathbf{B}(L^2, L^2)$-valued analytic in $z \in D_\epsilon$. Without loss of generality, we take $\gamma = (1, 0, \cdots, 0)$, and write $x \in \mathbf{R}^n$ as $x = (x_1, x')$, and let $\Delta' = \sum_{j=2}^{n}(\partial/\partial x_j)^2$. Recalling the expression of the Green kernel by the Hankel function, one sees that for any $\delta > 0$, $(-\Delta' - z)^{-1}$ defined on $\mathbf{C}_\pm$ has an analytic continuation across the positive real axis into the region $\{z \,;\, \pm\mathrm{Im}\, \sqrt{z} > -\delta\}$ on the space of functions of exponential weight. Denoting this operator by $r_\pm(z)$, we define

$$W_{\gamma,0}(E, z) = (\mathscr{F}_{x_1 \to \xi_1})^{-1}\big\{r_+(E - (\xi_1 + z)^2)\chi(\xi_1 < 0)$$

$$+ r_-(E - (\xi_1 + z)^2)\chi(\xi_1 > 0)\big\}\varphi_0(\xi_1)\mathscr{F}_{x_1 \to \xi_1}.$$

For $a \in \mathbf{R}$, define the space $\mathscr{H}_a$ by

$$\mathscr{H}_a = \left\{f \,;\, \int_{\mathbf{R}^n} e^{2a|x|}|f(x)|^2 dx < \infty\right\}$$

with obvious norm.

Theorem 3.10 *Let $E > 0$.*

(1) For any $\delta > 0$, there exists $\epsilon > 0$ such that $U_{\gamma,0}(E, z)$ is a $\mathbf{B}(\mathscr{H}_\delta, \mathscr{H}_{-\delta})$-valued analytic function on D_ϵ.

(2) $U_{\gamma,0}(E, z)$ has a continuous boundary value for $z \in \overline{D_\epsilon \cap \mathbf{R}}$, and for $t \in (-\epsilon/2, \epsilon/2)$

$$U_{\gamma,0}(E, t) = G_{\gamma,0}(\sqrt{E - t^2}, t).$$

(3) For $\tau > 0$

$$U_{\gamma,0}(E, i\tau) = G_{\gamma,0}(\sqrt{E + \tau^2}, i\tau).$$

(4) For $0 < s < 1$

$$\|U_{\gamma,0}(E, i\tau)\|_{\mathbf{B}(L^{2,s};L^{2,s-1})} \le C/\tau, \quad \tau > 1.$$

(5) Let

$$R_{\gamma,0}(E,t) = e^{it\gamma \cdot x} U_{\gamma,0}(E,t) e^{-it\gamma \cdot x}. \tag{3.24}$$

Then

$$(-\Delta - E) R_{\gamma,0}(E,t) = 1.$$

For the proof see [87]. The above theorem shows that the Faddeev Green operator on the imaginary axis $G_{\gamma,0}(\sqrt{E + \tau^2}, i\tau)$ is obtained as the analytic continuation of that on the real axis $G_{\gamma,0}(\sqrt{E - t^2}, t)$.

3.8 Inverse Scattering at a Fixed Energy

A remarkable fact found in the study of multi-dimensional inverse scattering theory is the reconstruction of the potential from the scattering matrix of one fixed energy. Assume that the potential is exponentially decreasing, i.e.

$$|V(x)| \leq C e^{-\delta_0 |x|} \tag{3.25}$$

for some constants $C, \delta_0 > 0$. Take $E > 0$ arbitrarily. Then for $0 < \delta < \delta_0/2$, $U_{\gamma,0}(E,z)V$ is compact on $\mathscr{H}_{-\delta}$. We define the set of *exceptional points* $\mathscr{E}_\gamma(E)$ to be the set of $z \in \overline{D_\epsilon}$ such that $-1 \in \sigma_p(U_{\gamma,0}(E,z)V)$. This is shown to be independent of δ. The following lemma can be proven by using the analytic Fredholm theorem (see, e.g., [167], Vol. 1, p. 211) and Fatou's theorem for boundary values of analytic functions (see, e.g., [116], p. 57).

Lemma 3.5 $\mathscr{E}_\gamma(E) \cap \mathbf{C}_+$ *is discrete and there exists a constant* $C > 0$ *such that* $\{i\tau \in \mathscr{E}_\gamma(E) ; \tau > C\} = \emptyset$. *Moreover,* $\mathscr{E}_\gamma(E) \cap \mathbf{R}$ *is a closed set of measure 0.*

By (3.24), for real t, $t \in \mathscr{E}_\gamma(E)$ is equivalent to $-1 \in \sigma_p(R_{\gamma,0}(E,t)V)$. We then define for $E > 0$ and $t \in (-\epsilon/2, \epsilon/2) \setminus \mathscr{E}_\gamma(E)$

$$R_\gamma(E,t) = (1 + R_{\gamma,0}(E,t)V)^{-1} R_{\gamma,0}(E,t).$$

Recall that the *physical scattering amplitude* $A(E)$ is defined by

$$\widehat{S}(E) = 1 - 2\pi i\, A(E).$$

It has the following representation

$$A(E) = \mathscr{F}_0(E)(V - VR(E+i0)V)\mathscr{F}_0(E)^*.$$

We replace the physical Green operator $R(E + i0)$ by the Faddeev Green operator $R_\gamma(E, t)$, and define the *Faddeev scattering amplitude* by

$$A_\gamma(E, t) = \mathscr{F}_0(E)(V - V R_\gamma(E, t)V)\mathscr{F}_0(E)^*.$$

It follows from (3.23) that $R_{\gamma,0}(E, t)$ satisfies

$$R_{\gamma,0}(E, t) = R_0(E + i0) - T_\gamma,$$

$$T_\gamma = 2\pi i \mathscr{F}_0(E)^* \chi_\gamma(t)\mathscr{F}_0(E), \quad \chi_\gamma(t) = \chi\left(\gamma \cdot \theta > \frac{t}{\sqrt{E}}\right).$$

Therefore, by Theorem 3.7, we have the following equation

$$A_\gamma(E, t) = A(E) + 2\pi i A(E)\chi_\gamma(t)A_\gamma(E, t). \tag{3.26}$$

For $t \in (-\epsilon/2, \epsilon/2) \setminus \mathscr{E}_\gamma(E)$, the perturbed Green operator $R_\gamma(E, t)$ exists. Therefore, by virtue of Theorem 3.8, the Eq. (3.26) is solvable with respect to $A_\gamma(E, t)$:

$$A_\gamma(E, t) = (1 - 2\pi i A(E)\chi_\gamma(t))^{-1}A(E)$$

for $t \in (-\epsilon/2, \epsilon/2) \setminus \mathscr{E}_\gamma(E)$. One can thus construct the Faddeev scattering amplitude from the physical scattering amplitude.

Theorem 3.11 *Let $n \geq 3$, and suppose $V(x)$ satisfy (3.25). Then, one can reconstruct $V(x)$ uniquely from the scattering matrix of arbitrarily fixed energy $E > 0$.*

Proof We construct the Faddeev scattering amplitude $A_\gamma(E, t)$ from the physical scattering amplitude $A(E)$. Up to a constant, it is written as

$$\int e^{-i\sqrt{E}(\theta-\theta')\cdot x}V(x)dx - \int e^{-i\sqrt{E}\theta\cdot x}V(x)R_\gamma(E, t)(V(\cdot)e^{i\sqrt{E}\theta'})dx.$$

Take $\sqrt{E}\theta = \sqrt{E - t^2}\omega + t\gamma$, $\sqrt{E}\theta' = \sqrt{E - t^2}\omega' + t\gamma$, where $\omega, \omega' \in S^{n-1}$ and $\omega \cdot \gamma = \omega' \cdot \gamma = 0$. Then the above kernel has the form

$$B_\gamma(\omega, \omega', t) = \int e^{-i\sqrt{E-t^2}(\omega-\omega')\cdot x}V(x)dx$$

$$- \int e^{-i\sqrt{E-t^2}\omega\cdot x}V(x)U_\gamma(E, t)(V(\cdot)e^{i\sqrt{E-t^2}\omega\cdot})dx,$$

where

$$U_\gamma(E,t) = e^{-it\gamma \cdot x} R_\gamma(E,t) e^{it\gamma \cdot x} = (1 + U_{\gamma,0}(E,t)V)^{-1} U_{\gamma,0}(E,t).$$

Note that $U_\gamma(E,t)$ has a unique meromorphic extension to D_ϵ and

$$\|U_\gamma(E,i\tau)\|_{\mathbf{B}(L^{2,s};L^{2,s-1})} \le C_s/\tau, \quad 0 < s < 1$$

for large $\tau > 0$. Therefore, as $\tau \to \infty$

$$B_\gamma(\omega,\omega',i\tau) \sim \int e^{-i\sqrt{E+\tau^2}(\omega-\omega') \cdot x} V(x) dx.$$

Here, we use the assumption $n \ge 3$. For any $\xi \in \mathbf{R}^n$, take $\gamma, \eta \in S^{n-1}$ such that $\xi \cdot \gamma = \xi \cdot \eta = \gamma \cdot \eta = 0$, and put

$$\omega = \omega(\tau) = (1 - \frac{\xi^2}{4\tau^2})^{1/2}\eta + \frac{\xi}{2\tau}, \quad \omega' = \omega'(\tau) = (1 - \frac{\xi^2}{4\tau^2})^{1/2}\eta - \frac{\xi}{2\tau}.$$

Then, $\sqrt{E+\tau^2}(\omega - \omega') \to \xi$. Therefore $B_\gamma(\omega(\tau), \omega'(\tau), i\tau) \to \widehat{V}(\xi)$.

When the potential is not exponentially decreasing, it cannot be determined from the scattering amplitude of a fixed energy (see [168]). However, one can construct $\widehat{V}(\xi)$ for $|\xi| < 2\sqrt{b}$ from the scattering amplitude for the energy interval (a,b). The following theorem can be proven by using the Faddeev Green operator instead of Eskin-Ralston Green operator. See [111] and [86].

Theorem 3.12 *Let $n \ge 2$, and suppose $V(x)$ satisfies*

$$|\partial_x^\alpha V(x)| \le C(1+|x|)^{-3/2-\epsilon-|\alpha|}, \quad |\alpha| \le sn - 1.$$

Let I be a set of positive measure on $\mathbf{R}$ such that $b = \text{ess.sup } I > 0$. Suppose we are given the scattering amplitude $A(E)$ for all $E \in I$. Then, we can reconstruct $\widehat{V}(\xi)$ for all $|\xi| < 2\sqrt{b}$. If I is a half-line: $I = [E_0, \infty)$, for any $\omega, \omega' \in S^{n-1}$ such that $\omega \cdot \gamma = \omega \cdot \gamma' = 0$ and $\omega \neq \omega'$, and for any $\lambda > 0$, one can construct a function $C_\gamma(\lambda, t, \omega, \omega')$ from the scattering matrix by which

$$\widehat{V}(\sqrt{\lambda}(\omega - \omega')) = C_\gamma(\lambda, t_0, \omega, \omega') + \text{p.v.} \frac{1}{\pi} \int_{-\infty}^{\infty} \frac{C_\gamma(\lambda, t, \omega, \omega')}{t - t_0} dt$$

holds for all $t_0 \in \mathbf{R}$.

3.9 Multi-Dimensional Gel'fand–Levitan Theory

Let us observe Faddeev theory more closely.

3.9.1 Justification of Volterra Properties

By Theorem 3.10, $R_{\gamma,0}(E,t) = \widetilde{R}_{\gamma,0}(E,t)$ when $E = \lambda^2 + t^2$ and $|t| < \epsilon/2$. Therefore, we redefine $R_{\gamma,0}(E,t)$ for $|t| \geq \epsilon/2$ by this formula. Define the perturbed Green operator by

$$R_\gamma(E,t) = (1 + R_{\gamma,0}(E,t)V)^{-1} R_{\gamma,0}(E,t).$$

For the potential satisfying $|V(x)| \leq C(1+|x|)^{-1-\epsilon_0}$ with $\epsilon_0 > 0$, $R_{\gamma,0}(E,t)V = e^{it\gamma \cdot x}G_{\gamma,0}(\lambda,t)Ve^{-it\gamma \cdot x}$ is compact on $L^{2,-s}$ for $1/2 < s < (1+\epsilon_0/2)$. Define the set of exceptional points $\widetilde{\mathscr{E}}_\gamma(\lambda)$ by

$$\widetilde{\mathscr{E}}_\gamma(\lambda) = \{z \in \overline{\mathbf{C}_+} \,;\; -1 \in \sigma_p(G_{\gamma,0}(\lambda,z)V)\}.$$

Lemma 3.6 *For $\lambda \geq 0$, $\widetilde{\mathscr{E}}_\gamma(\lambda) \cap \mathbf{C}_+$ is a discrete set, and $\widetilde{\mathscr{E}}_\gamma(\lambda) \cap \{|z| > C_0\} = \emptyset$ for large $C_0 > 0$. Moreover, $\widetilde{\mathscr{E}}_\gamma(\lambda) \cap \mathbf{R}$ is a closed set of measure zero.*

The existence of real exceptional points is a critical problem for the Faddeev theory. For small potentials they do not exist, however, Lavine and Nachman [127] and Khenkin and Novikov [111] proved that the real exceptional points do exist if $-\Delta + V$ has bound states.

Theorem 3.13 *If $\sigma_p(-\Delta + V) \neq \emptyset$, for any $\gamma \in S^{n-1}$, there exist $\lambda \geq 0$ and $t \in \mathbf{R}$ such that $-1 \in \sigma_p(G_{\gamma,0}(\lambda,t)V)$.*

We continue our arguments under the assumption that

$$-1 \notin \sigma_p(R_{\gamma,0}(\lambda,t)V), \quad \forall E > 0, \quad -\sqrt{E} \leq \forall t \leq \sqrt{E}.$$

This is satisfied when the potential is sufficiently small. We put

$$\Psi_\gamma(x,E,\theta) = e^{i\sqrt{E}\theta \cdot x} - R_\gamma(E, \sqrt{E}\theta \cdot \gamma)(V(\cdot)e^{i\sqrt{E}\theta \cdot}),$$

$$\widetilde{A}_\gamma(E,\theta,\theta') = (2\pi)^{-n}2^{-1}E^{(n-2)/2} \int_{\mathbf{R}^n} e^{-i\sqrt{E}\theta \cdot x}V(x)\Psi_\gamma(x,E,\theta')dx,$$

$$Q_\gamma^{(\pm)}(E,\theta,\theta') = 2\pi i\chi(\pm\gamma \cdot (\theta - \theta') \geq 0)\widetilde{A}_\gamma(E,\theta,\theta'),$$

and let $Q_\gamma^{(\pm)}(E)$ be the integral operator with kernel $Q_\gamma^{(\pm)}(E,\theta,\theta')$.

We can now prove the factorization formula of the S-matrix (3.15). Letting

$$L_\gamma^{(\pm)}(t) = 2\pi i \chi(\pm\gamma \cdot (\theta - \frac{t\gamma}{\sqrt{E}}) \geq 0)A_\gamma(E,t),$$

we have

$$\widehat{S}(E)(1 + L_\gamma^{(+)}(t)) = 1 - L_\gamma^{(-)}(t).$$

Observing its integral kernel, we have

$$\int_{S^{n-1}} \widehat{S}(E,\theta,\theta')(\delta(\theta''-\theta') + L_\gamma^{(+)}(t,\theta'',\theta'))d\theta'' = \delta(\theta-\theta') - L_\gamma^{(-)}(t,\theta,\theta').$$

Letting $t = \sqrt{E}\theta \cdot \gamma$ in the above formula, we get

$$\widehat{S}(E)(1 + Q_\gamma^{(+)}(E)) = 1 - Q_\gamma^{(-)}(E).$$

One can show that $(Q_\gamma^{(\pm)}(E))^2 = 0$, hence $1 + Q_\gamma^{(+)}(E)$ is invertible. This and the above formula prove (3.15).

We prove (3.16). As far as the author knows, the rigorous proof was not presented. Below, we show author's own proof, whose details must again be omitted due to the lack of space. Letting

$$\Phi_\gamma(x,\xi) = \Psi_\gamma(x, |\xi|^2, \xi/|\xi|),$$

we have

$$\Phi(x,\xi) = e^{ix\cdot\xi} - \widetilde{K}_\gamma(x,\xi),$$

$$\widetilde{K}_\gamma(x,\xi) = \widetilde{R}_\gamma(|\xi'|^2, \xi \cdot \gamma)(V(\cdot)e^{i\xi}), \quad \xi' = \xi - (\xi \cdot \gamma)\gamma.$$

Here, $\widetilde{R}_\gamma(\lambda, t)$ is defined by

$$\widetilde{R}_\gamma(\lambda, t) = (1 + \widetilde{R}_{\gamma,0}(\lambda, t)V)^{-1}\widetilde{R}_{\gamma,0}(\lambda, t).$$

We put $\gamma = (1, 0, \cdots, 0)$ for the sake of simplicity. Assume that $n \geq 3$ and

$$|V(x)| \leq C(1 + |x|)^{-\rho}, \quad \rho > \max\{2, (n+1)/2\}.$$

Then, one can show that if $|\xi'| \geq \epsilon_0 > 0$

$$|\widetilde{K}_\gamma(x, \xi_1, \xi')| \leq C(1 + |\xi|)^{-1},$$

$$\left|\frac{\partial}{\partial\xi_1}\widetilde{K}_\gamma(x, \xi_1, \xi')\right| \leq C|\xi'|(1 + |\xi|)^{-1}.$$

We let

$$K_\gamma(x, y_1, \xi') = (2\pi)^{-1/2} \int_{-\infty}^{\infty} e^{-iy_1\xi_1} \widetilde{K}_\gamma(x, \xi_1, \xi')d\xi_1,$$

where the integral is taken in the limit in the mean. One can then show that

$$(1 + |y_1|)K_\gamma(x, y_1, \xi') \in L^2(\mathbf{R}_{y_1}),$$

which implies

$$K_\gamma(x, y_1, \xi') \in L^2(\mathbf{R}_{y_1}).$$

Noting that

$$K_\gamma(x, y_1, \xi') = (2\pi)^{-1/2} \int_{-\infty}^{\infty} e^{-i(y_1-x_1)\xi_1} G_\gamma(|\xi'|^2, \xi_1)(V(x)e^{ix'\cdot\xi'})d\xi_1,$$

and that $G_\gamma(\lambda, z)$ is analytic in $z \in \mathbf{C}_+$, we have by the Paley–Wiener theorem, $K_\gamma(x, y_1, \xi') = 0$ if $y_1 - x_1 < 0$. Therefore,

$$\Phi_\gamma(x, \xi) = e^{ix\cdot\xi} - (2\pi)^{-1/2} \int_{x_1}^{\infty} e^{iy_1\xi_1} K_\gamma(x, y_1, \xi')dy_1,$$

which proves (3.16).

3.9.2 Gel'fand–Levitan Equation

We no longer have sufficient estimates to guarantee the validity of remaining arguments. Therefore, the following arguments are formal, although quite intriguing.

We introduce three types of integral operators

$$T_0 f(\xi) = (2\pi)^{-n/2} \int_{\mathbf{R}^n} e^{-ix\cdot\xi} f(x)dx,$$

$$T f(\xi) = (2\pi)^{-n/2} \int_{\mathbf{R}^n} \overline{\Phi(x, \xi)} f(x)dx,$$

$$T_\gamma f(\xi) = (2\pi)^{-n/2} \int_{\mathbf{R}^n} \overline{\Phi_\gamma(x, \xi)} f(x)dx,$$

where $\Phi(x, \xi) = e^{ix \cdot \xi} - R(|\xi|^2 + i0)(V(\cdot)e^{-i\xi \cdot})$. We define

$$Q_\gamma^{(+)} f(\xi) = (Q_\gamma^{(+)}(|\xi|^2) f(|\xi| \cdot))(\xi / |\xi|).$$

Then, one can show

$$T^* = T_\gamma^* (1 - Q_\gamma^{(+)}).$$

Now, we can derive the Gel'fand–Levitan equation. For the sake of simplicity, we assume that $\sigma_p(-\Delta + V) = \emptyset$. In view of (3.16), we put

$$U_\gamma = T_\gamma^* T_0 = 1 + K_\gamma.$$

Putting

$$\tilde{Q}_\gamma = Q_\gamma^{(+)} Q_\gamma^{(+)*} - Q_\gamma^{(+)} - Q_\gamma^{(+)*},$$

and using $T^* T = 1$, we have

$$T_\gamma^* (1 + \tilde{Q}_\gamma) T_\gamma = 1,$$

Replacing T_γ by $T_0 U_\gamma^* = T_0(1 + K_\gamma^*)$, we have

$$1 + K_\gamma + \Omega_\gamma + K_\gamma \Omega_\gamma = (1 + K_\gamma^*)^{-1},$$

where

$$\Omega_\gamma = T_0^* \tilde{Q}_\gamma T_0.$$

Since $K_\gamma^*(x, y)$ is supported in $\{(x - y) \cdot \gamma \geq 0\}$, letting $(1 + K_\gamma^*)^{-1} = 1 + C_\gamma$, we see that C_γ is also supported in $\{(x - y) \cdot \gamma \geq 0\}$. Then, for $(x - y) \cdot \gamma < 0$, we have

$$K_\gamma(x, y) + \Omega_\gamma(x, y) + \int_{(x-y) \cdot \gamma < 0} K_\gamma(x, z) \Omega_\gamma(z, y) dz = 0.$$

This is the multi-dimensional Gel'fand–Levitan equation.

Little is known about the solvability of this equation. It might be solved for small scattering amplitudes to obtain K_γ. By differentiating it, we might get $V(x)$. However, its estimates (e.g., the spatial decay) would be hard to prove.

Melin [142] considered an approach based on the ultra-hyperbolic equation $(\Delta_x - \Delta_y)U = V(x)U$. His starting point is a fundamental solution to the free ultra-hyperbolic equation $(\Delta_x - \Delta_y)U = 0$, which has an advantage that it directly constructs the intertwining operator with triangular kernel.

3.10 $\overline{\partial}$-Approach

In the 1980s, $\overline{\partial}$-approach was introduced as a new viewpoint of the inverse scattering by Beals and Coifman [14] and Nachman and Ablowitz [150]. In this approach, Faddeev's method is rewritten as follows. Let $\widetilde{G}(\zeta)$ be as in (2.7). The key fact is the following equation. Let $\overline{\partial}_j = \partial/\partial\overline{\zeta}_j$.

Theorem 3.14 *For* $\operatorname{Im}\zeta_j \neq 0$,

$$\overline{\partial}_j \widetilde{G}(\zeta)f = (2\pi)^{1-n/2} \int_{\mathbf{R}^n} e^{ix\cdot\xi} \widehat{f}(\xi)\xi_j \delta(\xi^2 + 2\zeta\cdot\xi)d\xi,$$

and $\overline{\partial}_j \widetilde{G}(\zeta) \in \mathbf{B}(L^{2,s}; L^{2,-s})$, $s > 1$.

Faddeev scattering amplitude is then written as follows. Define

$$\widetilde{G}_V(\zeta) = (1 + \widetilde{G}(\zeta)V)^{-1}\widetilde{G}(\zeta).$$

If $i\tau \notin \mathscr{E}_\gamma(E)$, it satisfies

$$\widetilde{G}_V(\eta + i\tau\gamma) = e^{-ix\cdot\eta}U_\gamma(E, i\tau)e^{ix\cdot\eta}.$$

For $\omega, \omega' \in S^{n-1}$ satisfying $\omega\cdot\gamma = \omega'\cdot\gamma = 0$, we let $\eta = \sqrt{E + \tau^2}\omega$, $\eta' = \sqrt{E + \tau^2}\omega'$. Then,

$$B_\gamma(\omega, \omega', i\tau) = \int e^{-i(\eta-\eta')\cdot x}V(x)dx - \int e^{-i(\eta-\eta')\cdot x}V(x)\widetilde{G}_V(\zeta)V\,dx,$$

where $\zeta = \eta' + i\tau\gamma$. Then, $\xi = \eta - \eta'$ satisfies $\xi^2 + 2\zeta\cdot\xi = 0$. Starting from the scattering amplitude, we have constructed

$$T(\xi, \zeta) = \int e^{-ix\cdot\xi}V(x)dx - \int e^{-ix\cdot\xi}V(x)\widetilde{G}_V(\zeta)V\,dx,$$

on the set $\{(\xi, \zeta)\}$, where $\xi \in \mathbf{R}^n$ and $\zeta \in \mathbf{C}^n$ satisfy $\zeta^2 = E, |\zeta| > C, \operatorname{Im}\zeta \neq 0$ and $\xi^2 + 2\zeta\cdot\xi = 0$, C being a large constant. This set has a structure of fibered space and each fiber

$$\mathscr{V}_\xi = \{\zeta \in \mathbf{C}^n \,;\, \zeta^2 = E, |\zeta| > C, \operatorname{Im}\zeta \neq 0, \xi^2 + 2\zeta\cdot\xi = 0\}$$

is a complex manifold of dimension $2n - 4$. On this complex manifold, $T(\xi, \zeta)$ satisfies the following $\overline{\partial}$-equation. Let

$$A_j(\xi, \zeta) = -(2\pi)^{1-n/2} \int_{\mathbf{R}^n} T(\xi - \eta, \zeta + \eta)T(\eta, \zeta)\eta_j \delta(\eta^2 + 2\zeta\cdot\eta)d\eta.$$

Theorem 3.15 *As a 1-form on $\mathcal{V}_\xi$, we have*

$$\bar{\partial} T(\xi, \zeta) = \sum_{j=1}^{n} A_j(\xi, \zeta) d\bar{\zeta}_j.$$

There are two remarkable applications of this theorem. Recall the generalized Cauchy formula

$$f(z) = \frac{1}{2\pi i} \int_{\partial D} \frac{f(\zeta)}{\zeta - z} d\zeta - \frac{1}{2\pi i} \int_{\partial D} \frac{\bar{\partial} f(\zeta)}{\zeta - z} d\bar{\zeta} \wedge d\zeta,$$

where D is a domain in $\mathbf{C}$. Using a generalization of this formula on the complex manifold $\mathcal{V}_\xi$, Nachman [148] derived the following representation formula for the potential by means of $T(\xi, \zeta)$:

$$\widehat{V}(\xi) = T(\xi, \zeta_0) + \int_{\partial \mathcal{V}_\xi} T(\xi, \zeta) K(\zeta, \zeta_0) + \int_{\mathcal{V}_\xi} \sum_{j=1}^{n} A_j(\xi, \zeta) d\bar{\zeta}_j \wedge K(\zeta, \zeta_0),$$

where $\zeta_0 \in \mathcal{V}_\xi$ and $K(\zeta, \zeta_0)$ is a suitable $2n - 5$ form on $\mathcal{V}_\xi$.

Another important application is the characterization of the Faddeev scattering amplitude. The $\bar{\partial}$-equation in Theorem 3.15 gives a necessary and sufficient condition for a function $T(\xi, \zeta)$ defined on $\mathcal{V}_\xi$ to be the Faddeev scattering amplitude associated with some Schrödinger operator $-\Delta + V(x)$. This fact was discussed by Beals and Coifman [14] and Khenkin and Novikov [111]. The advantage compared to Faddeev's characterization is that in the $\bar{\partial}$-approach, one can get precise estimates of $T(\xi, \zeta)$ and also specify the associated class of potentials. However, the characterization of the physical scattering amplitude is still open, although it is linked with the Faddeev scattering amplitude through the Eq. (3.26).

Chapter 4
Boundary Control Method

As we have seen above, the theory of Gel'fand–Levitan–Marchenko deals with stationary solutions of Schrödinger equations, and uses analytic properties with respect to energy of its spectral data. In the early stage of the study of one-dimensional inverse problems, M. G. Krein proposed an approach to inverse problems for the wave equation [118, 119]. In contrast to Gel'fand–Levitan–Marchenko's theory, Krein's idea is based on the finite propagation property of the wave equation in space-time. However, this distinguished feature was disguised by its stationary formulation and the analytic properties of Fourier transforms of solutions. Blagovestcenskii [23, 24] made this time-dependent, hyperbolic nature more apparent. The crucial step was to introduce an identity which relates the spectral data with the solution of the boundary value problem. These ideas were extended to the multi-dimensional case by Belishev [15] using the controllability of solutions, and then to Riemannian manifolds by Belishev–Kurylev [16]. Since the wave propagates along the geodesic, the wave motion contains much information of the Riemannian metric and makes it possible to recover the metric from solutions of the boundary value problem, hence from the spectral data. This inverse spectral procedure is called boundary control method (BC-method), which we explain in this chapter. The BC-method is a rather long procedure. Therefore, we only give its outline.

The key idea of the BC-method is to make a copy of the manifold $\mathcal{M}$ in a space of functions on the boundary $\partial\mathcal{M}$.[1] We start from the spectral theory. In Sect. 4.1, we make a bridge from the boundary spectral data (BSD) to the function space over the support of solutions for the boundary value problem of the time-dependent wave equation. Then, the main geometric role is played by the boundary distance function $r_x(z) = d(x, z), x \in \mathcal{M}, z \in \partial\mathcal{M}$, to be introduced in Sect. 4.2, which induces a map $\mathcal{R} : \mathcal{M} \to C(\partial\mathcal{M})$. From $\mathcal{R}$, we reconstruct the topology, differentiable

---

[1] A general idea is that the space is determined by functions defined on it.

© The Author(s), under exclusive licence to Springer Nature Singapore Pte Ltd. 2020
H. Isozaki, *Inverse Spectral and Scattering Theory*, SpringerBriefs
in Mathematical Physics 38, https://doi.org/10.1007/978-981-15-8199-1_4

structure, and the Riemannian metric of $\mathcal{M}$. The reconstruction scheme is then as follows:

$$BSD \implies \mathcal{R}(\mathcal{M}) \implies \begin{cases} topology \\ differentiable\ structure \\ metric. \end{cases}$$

To implement this procedure, the theory of PDE, in particular the uniqueness of solutions across a non-characteristic surface, plays a key role in guaranteeing the controllability of solutions to the wave equation. In Sect. 4.9, we show an application of BC-method to the inverse scattering on non-compact manifolds.

4.1 The Role of Spectral Theory

Let $\mathcal{M}$ be an n-dimensional complete connected Riemannian manifold with boundary $\partial\mathcal{M}$. Let $ds^2 = \sum_{i,j=1}^{n} g_{ij}(x)dx^i dx^j$ be its Riemannian metric, $g = \det\left(g_{ij}\right)$ and $\left(g^{ij}\right) = \left(g_{ij}\right)^{-1}$. Letting $\partial_i = \partial/\partial x^i$, the Laplace operator on $\mathcal{M}$ is written as $\Delta_g = \frac{1}{\sqrt{g}} \sum_{i,j=1}^{n} \partial_i \left(\sqrt{g}\, g^{ij} \partial_j\right)$. Let $H^m(\mathcal{M})$ be the Sobolev space of order m on $\mathcal{M}$. Note that the inner product of $L^2(\mathcal{M})$ and $H^1(\mathcal{M})$ are defined by[2]

$$(u, v)_{L^2(\mathcal{M})} = \int_{\mathcal{M}} u(x)\overline{v(x)}dV_g, \quad dV_g = \sqrt{g(x)}dx,$$

$$(u, v)_{H^1(\mathcal{M})} = \sum_{i,j=1}^{n} \int_{\mathcal{M}} g^{ij}\partial_i u \overline{\partial_j v} dV_g + (u, v)_{L^2(\mathcal{M})}.$$

Let L^N be the Neumann Laplacian:

$$L^N u = -\Delta_g u, \quad u \in H^2(\mathcal{M}), \quad \partial_\nu u\Big|_{\partial\mathcal{M}} = 0.$$

Let $\lambda_k, \varphi_k, k = 1, 2, 3, \cdots$, be the associated eigenvalues and normalized eigenvectors:

$$0 = \lambda_1 < \lambda_2 \leq \lambda_3 \leq \cdots \to \infty, \quad -\Delta_g \varphi_k = \lambda_k \varphi_k, \quad \partial_\nu \varphi_k\Big|_{\partial\mathcal{M}} = 0,$$

---

[2] We need a suitable partition of unity on $\mathcal{M}$.

and call $\{(\lambda_k, \varphi_k|_{\partial\mathcal{M}})\}_{k=1}^{\infty}$ the *boundary spectral data* (BSD). Then, the problem we address is the following:

Question: Given BSD, can we determine $(\mathcal{M}, g)$?

More precisely, given an $(n-1)$-dimensional compact manifold S, a set of non-negative numbers $\{\lambda_k\}_{k=1}^{\infty}$ and functions $\{\psi_k\}_{k=1}^{\infty}$ on S, we try to construct a Riemannian manifold $\mathcal{M}$ whose boundary is S and its Laplace operator has $\{\lambda_k\}_{k=1}^{\infty}$ as Neumann eigenvalues. The best solution, characterization, is hard to answer. Therefore, we admit that such a Riemannian manifold exists and reconstruct the manifold and its metric from the knowledge of BSD.

We consider an initial-boundary value problem (IBVP) for the wave equation

$$
\begin{cases}
\partial_t^2 u = \Delta_g u & \text{on} \quad \mathcal{M} \times (0, \infty), \\[2mm]
u\big|_{t=0} = \partial_t u\big|_{t=0} = 0 & \text{on} \quad \mathcal{M}, \\[2mm]
\partial_\nu u\big|_{\partial\mathcal{M} \times (0,\infty)} = f \in C_0^\infty(\partial M \times (0, \infty)),
\end{cases}
\tag{4.1}
$$

where $\nu = (\nu_1, \cdots, \nu_n)$ is the outer unit normal vector to $\partial\mathcal{M}$ and $\partial_\nu = \sum_{i,j=1}^{n} \nu_i g^{ij} \partial_j$ on $\partial\mathcal{M}$. It is well-known that (4.1) has a unique solution $u^f(x, t)$. Expanding it by eigenvectors:

$$
u^f(x, t) = \sum_k u_k^f(t) \varphi_k(x), \quad u_k^f(t) = \int_{\mathcal{M}} u^f(y, t) \varphi_k(y) dV_g,
$$

and using the Parseval's formula, we have the following lemma.

Lemma 4.1 $\left(u^f(t), u^h(s)\right)_{L^2(\mathcal{M})} = \sum_k u_k^f(t) \overline{u_k^h(s)}.$

Although this lemma is elementary, it is a starting point of BC-method. We look at this formula from a different viewpoint. By the equation (4.1), we have[3]

$$
\frac{d^2}{dt^2} u_k^f(t) + \lambda_k u_k^f(t) = \int_{\partial\mathcal{M}} f(y, t) \varphi_k(y) dS_g, \quad u_k^f(0) = \frac{d}{dt} u_k^f(0) = 0.
$$

Solving this equation, we obtain

$$
u_k^f(t) - \int_0^t ds \int_{\partial\mathcal{M}} dS_g \frac{\sin(\sqrt{\lambda_k}(t-s))}{\sqrt{\lambda_k}} f(y, s) \varphi_k(y).
$$

---

[3] Apply Green's formula to $(\partial_t^2 u^f(t), \varphi_k)_{L^2(\mathcal{M})} = (\Delta_g u^f(t), \varphi_k)_{L^2(\mathcal{M})}.$

Noting that the right-hand side depends only on BSD, we arrive at the first step of BC-method.

Corollary 4.1 *Given the knowledge of $\partial \mathcal{M}$ and the surface element on it, one can compute $(u^f(t), u^h(t))_{L^2(\mathcal{M})}$ from BSD.*

4.2 Boundary Distance Function: From $\mathcal{R}(\mathcal{M})$ to $\mathcal{M}$

Let $d(x, y)$ be the distance of $x, y \in \mathcal{M}$, which is defined to be the infimum of length of piecewise smooth geodesic between x and y. The *boundary distance function r_x* and the map $\mathcal{R}$ are defined by

$$r_x(z) = d(x, z), \quad x \in \mathcal{M}, \quad z \in \partial \mathcal{M},$$

$$\mathcal{R} : \mathcal{M} \ni x \to r_x(\cdot) \in C(\partial \mathcal{M}).$$

If $\partial \mathcal{M}$ is compact, $\mathcal{R}(\mathcal{M}) = \{r_x ; x \in \mathcal{M}\}$ becomes a metric space by the distance

$$d_\infty(r_x, r_y) = \|r_x(\cdot) - r_y(\cdot)\|_{L^\infty(\partial \mathcal{M})},$$

and the following inclusion relations hold

$$\mathcal{R}(\mathcal{M}) \subset C^{0,1}(\partial \mathcal{M}) \subset L^\infty(\partial \mathcal{M}),$$

where $C^{0,1}(\partial \mathcal{M})$ is the set of Lipschitz continuous functions on $\partial \mathcal{M}$. One can show that the map $\mathcal{R} : (\mathcal{M}, d) \to (R(\mathcal{M}), d_\infty)$ is continuous and injective. By a theorem from general topology, this implies the following lemma.

Lemma 4.2 *If $\partial \mathcal{M}$ is compact, $(\mathcal{R}(\mathcal{M}), d_\infty)$ is homeomorphic to $(\mathcal{M}, d)$.*

This lemma shows that the topology of $\mathcal{M}$ is determined by $\mathcal{R}(\mathcal{M})$.

4.3 From BSD to $\mathcal{R}(\mathcal{M})$

Our next aim is to prove that if two manifolds $\mathcal{M}^{(1)}$ and $\mathcal{M}^{(2)}$ have the same BSD, the spaces of boundary distance functions $\mathcal{R}(\mathcal{M}^{(1)})$ and $\mathcal{R}(\mathcal{M}^{(2)})$ coincide. If this is shown, by virtue of Lemma 4.2, the BSD determines the topology of $\mathcal{M}$.

4.3.1 From BSD to the Domain of Influence

For a subset $A \subset \mathcal{M}$, its *domain of influence* at time $\tau > 0$ is defined by

$$\mathcal{M}(A, \tau) = \{x \in \mathcal{M} \; ; \; d(x, A) \leq \tau\}.$$

For a subset $\Gamma \subset \partial\mathcal{M}$, we let $\chi_{\mathcal{M}(\Gamma, \tau)}(x)$ be the characteristic function of $\mathcal{M}(\Gamma, \tau)$, and define the projection $P_{\Gamma, \tau}$ on $L^2(\mathcal{M})$ by

$$P_{\Gamma, \tau} f(x) = \chi_{\mathcal{M}(\Gamma, \tau)}(x) f(x), \quad f \in L^2(\mathcal{M}).$$

For $z \in \partial\mathcal{M}$, we put

$$\mathcal{M}(z, \tau) = \mathcal{M}(\{z\}, \tau) = \{x \in \mathcal{M} \; ; \; d(x, z) \leq \tau\}, \quad P_{z, \tau} = P_{\{z\}, \tau}.$$

To determine the range of $P_{z, \tau}$ is a crucial step toward the reconstruction of the manifold $\mathcal{M}$. For this purpose, let us recall the projection theorem in functional analysis. Let S be a closed subspace of a Hilbert space $\mathcal{H}$. Then, for any $f \in \mathcal{H}$, there exists a unique $u \in S$ such that $\|f - u\| = \inf_{w \in S} \|f - w\|$. Moreover, $f - u \in S^{\perp}$, and for any sequence $\{w_n\}_{n=1}^{\infty} \subset S$ satisfying $\inf_{w \in S} \|f - w\| = \lim_{n \to \infty} \|f - w_n\|$, we have $\|w_n - u\| \to 0$.

We use this fact in the following way. Let $f \in C_0^{\infty}(\partial\mathcal{M} \times (0, \infty))$, $t, \tau > 0$, and $\Gamma \subset \partial\mathcal{M}$ be an open set. Then, we have

$$\|(1 - P_{\Gamma, \tau}) u^f(t)\| = \inf_{\eta \in C_0^{\infty}(\Gamma \times (0, \tau))} \|u^f(t) - u^{\eta}(\tau)\|. \tag{4.2}$$

Hence there exists a sequence $f_j \in C_0^{\infty}(\Gamma \times (0, \tau))$ satisfying

$$\|P_{\Gamma, \tau} u^f(t) - u^{f_j}(\tau)\| \to 0.$$

This procedure is not constructive, since we do not know f_j unless we know $\mathcal{M}$ and its metric. However, its feature is that this process is described by the inner product, hence relies only on BSD. Here, it is convenient to introduce the following expression. Let $A^{(1)}$ and $A^{(2)}$ be some analytical or geometrical quantities (formulas) associated with two manifolds $\mathcal{M}^{(1)}$ and $\mathcal{M}^{(2)}$. If $A^{(1)} = A^{(2)}$ whenever $\mathcal{M}^{(1)}$ and $\mathcal{M}^{(2)}$ have the same BSD, we say that *BSD determines A*. Then, we have the following lemma.

Lemma 4.3 *Let $f, h \in C_0^{\infty}(\partial\mathcal{M} \times (0, \infty))$ and $\tau_1, \tau_2, t, s > 0$.*

(1) Let $\Gamma_1, \Gamma_2 \subset \partial\mathcal{M}$ be open sets. Then BSD determines the inner product

$$(P_{\Gamma_1, \tau_1} u^f(t), P_{\Gamma_2, \tau_2} u^h(s))_{L^2(\mathcal{M})}.$$

(2) *Let $z_1, z_2 \in \partial \mathcal{M}$. Then BSD determines the inner product*

$$(P_{z_1, \tau_1} u^f(t), P_{z_2, \tau_2} u^h(s))_{L^2(\mathcal{M})}.$$

In fact, the assertion (2) follows from (1) by choosing open sets $\Gamma_1, \Gamma_2 \subset \partial \mathcal{M}$ shrinking to z_1, z_2.

We note here that in deriving (4.2), we need the following fact from the boundary value problem.

Theorem 4.1 *Define the map $\mathscr{C}_\tau^\Gamma$ by*

$$\mathscr{C}_\tau^\Gamma : L^2(\Gamma \times (0, \tau)) \ni f \to u^f \Big|_{t=\tau} \in L^2(\mathcal{M}(\Gamma, \tau)).$$

Then, we have

$$\overline{\mathrm{Ran}(\mathscr{C}_\tau^\Gamma)} = L^2(\mathcal{M}(\Gamma, \tau)).$$

This is a crucial step in the BC-method from the viewpoint of the boundary value problem for PDE. Granting it for the moment, we continue the reconstruction procedure of the manifold and metric.

4.3.2   Boundary Normal Geodesic

For a $z \in \partial \mathcal{M}$, a *boundary normal geodesic* $\gamma_z(t)$ emanating from z is a geodesic in $\mathcal{M}$ with initial point z having the unit normal at z as initial direction. To make it explicit, take local coordinates $z = (z^1, \cdots, z^{n-1})$ on $\partial \mathcal{M}$ and $x = (z^1, \cdots, z^{n-1}, x^n)$ as local coordinates near $\partial \mathcal{M}$, where $\partial \mathcal{M}$ is represented as $x^n = 0$. Then, the geodesic is the solution of the equation

$$\begin{cases} \dfrac{d^2 x^k}{dt^2} + \Gamma_{ij}^k(x(t)) \dfrac{dx^i}{dt} \dfrac{dx^j}{dt} = 0, \\[2mm] x(0) = (z, 0), \quad \dfrac{dx}{dt}(0) = \nu(z), \end{cases}$$

where $\nu(z)$ is the unit normal at z, and Γ_{ij}^k is the Christoffel symbol. The map $\gamma_z(t)$: $(z, t) \to x(t, z)$ is a diffeomorphism near $\partial \mathcal{M}$, and (z, t) is said to be a *boundary normal coordinate* near $\partial \mathcal{M}$.

For $x \in \mathcal{M}(\Gamma, \tau)$, there is a boundary normal geodesic $\gamma_z(\cdot)$ such that $\gamma_z(s) = x$ for some $s \leq \tau$. If $d(\gamma_z(s), z) = d(\gamma_z(s), \partial \mathcal{M})$, then $s = d(x, \partial \mathcal{M})$, which is equal to $d(x, z) = r_x(z)$, the boundary distance function evaluated at z. Therefore, to see whether or not $s = d(x, \partial \mathcal{M})$ is a key step toward the reconstruction of the boundary distance function. It is possible by using BSD due to the following two lemmas.

Lemma 4.4 *Let $s > 0$, and $\gamma_z(\cdot)$ be the boundary normal geodesic starting from $z \in \partial\mathcal{M}$. Then, the following 3 assertions are equivalent.*

(1) $d(\gamma_z(s), z) = d(\gamma_z(s), \partial\mathcal{M})$.
(2) For any $\epsilon > 0$ and any neighborhood $\Gamma \subset \partial\mathcal{M}$ of z, the interior of $\left(\mathcal{M}(\Gamma, s) \setminus \mathcal{M}(\partial\mathcal{M}, s - \epsilon)\right)$ is non-empty.
(3) For any $\epsilon > 0$ and any neighborhood $\Gamma \subset \partial\mathcal{M}$ of z, there exists $h \in C_0^\infty(\Gamma \times (0, s))$ such that $\|u^h(s)\| > \|P_{\partial\mathcal{M}, s-\epsilon} u^h(s)\|$.

Lemma 4.5 *Let $\gamma_w(\cdot)$ be the boundary normal geodesic starting from $w \in \partial\mathcal{M}$, and $s > 0$ be such that $d(\gamma_w(s), w) = d(\gamma_w(s), \partial\mathcal{M})$. Let $z \in \partial\mathcal{M}$ and $t > 0$. Then, the following 3 assertions are equivalent.*

(1) $t > d(\gamma_w(s), z)$.
(2) There exist a neighborhood $\Gamma \subset \partial\mathcal{M}$ of w and $\epsilon > 0$ such that

$$\mathcal{M}(\Gamma, s) \subset \mathcal{M}(\partial\mathcal{M}, s - \epsilon) \cup \mathcal{M}(z, t - \epsilon).$$

(3) There exist a neighborhood $\Gamma \subset \partial\mathcal{M}$ of w and $\epsilon > 0$ such that for any $h \in C_0^\infty(\Gamma \times (0, s))$

$$\|u^h(s)\|^2 = \|P_{\partial\mathcal{M}, s-\epsilon} u^h(s)\|^2 + \|P_{z, t-\epsilon} u^h(s)\|^2 - (P_{\partial\mathcal{M}, s-\epsilon} u^h(s), P_{z, t-\epsilon} u^h(s)).$$

In order to elucidate the role of PDE, let us prove Lemma 4.4 (3). Let χ be the characteristic function of $\mathcal{M}(\Gamma, s) \setminus \mathcal{M}(\partial\mathcal{M}, s - \epsilon)$. Then, by Lemma 4.4 (2), $\|\chi\|_{L^2(\mathcal{M})} > 0$. By virtue of Theorem 4.1, χ is approximated by $u^h(s)$ for some $h \in C_0^\infty(\Gamma \times (0, s))$, which proves Lemma 4.4 (3).

4.4 From the Domain of Influence to the Topology of $\mathcal{M}$

Let $\gamma_z(\cdot)$ be the boundary normal geodesic starting from $z \in \partial\mathcal{M}$. A point $\gamma_z(t)$ is said to be *uniquely minimizing* along the geodesic $\gamma_z(\cdot)$ if $t = d(\gamma_z(t), \partial\mathcal{M})$ and $t < d(\gamma_z(t), w)$ for any $w \in \partial\mathcal{M}$ such that $w \neq z$. Hence, $\{\gamma_z(s) ; 0 \leq s \leq t\}$ is a unique shortest geodesic from $\partial\mathcal{M}$ to $\gamma_z(t)$. We can then show that if $\gamma_z(t)$ is uniquely minimizing along $\gamma_z(\cdot)$, then so is $\gamma_z(s)$ for any $0 < s < t$. We put

$$\tau(z) = \sup\{t ; \gamma_z(t) \text{ is uniquely minimizing}\}$$

and call it *boundary cut function*. The point $\gamma_z(\tau(z))$ is called *boundary cut point* of z along $\gamma_z(\cdot)$. The *boundary cut locus* is defined by

$$\omega = \{\gamma_z(\tau(z)) ; z \in \partial\mathcal{M}\}.$$

Now, by virtue of Lemma 4.3 and Lemma 4.4 (3), BSD determines whether or not $\gamma_w([0, s])$ is a shortest geodesic to $\partial\mathcal{M}$, which then determines the boundary cut function $\tau(w)$ as well. By Lemma 4.5, for $s < \tau(w)$, BSD determines $d(\gamma_w(s), z)$ for any $z \in \partial\mathcal{M}$. Therefore, for any $w \in \partial\mathcal{M}$ and $s \le \tau(w)$, BSD determines $r^{(w,s)}(\cdot) \in C(\partial\mathcal{M})$ by

$$r^{(w,s)}(z) = d(\gamma_w(s), z), \quad z \in \partial\mathcal{M},$$

which turns out to be the boundary distance function corresponding to $x = \gamma_w(s)$. We have thus seen that BSD determines the boundary distance function $d(x, z)$.

Lemma 4.6 *We put*

$$\mathcal{B}(\mathcal{M}) = \cup_{z \in \partial\mathcal{M}}\{\gamma_z(t) \,;\, 0 \le t < \tau(z)\}.$$

(1) $\mathcal{B}(\mathcal{M})$ is an open set, and

$$\mathcal{M} = \mathcal{B}(\mathcal{M}) \cup \omega, \quad \mathcal{B}(\mathcal{M}) \cap \omega = \emptyset.$$

(2) ω is a closed set of measure 0. In particular, it has no interior points.

Due to Lemma 4.6, when (w, s) varies over $\partial\mathcal{M} \times [0, \tau(w)]$, $r^{(w,s)}(z)$ varies over whole $\mathcal{R}(\mathcal{M})$. We have thus seen that BSD determines $\mathcal{R}(\mathcal{M})$.

4.5   From $\mathcal{R}(\mathcal{M})$ to the Differentiable Structure

Once one knows $\mathcal{R}(\mathcal{M})$, one can introduce a differentiable structure on it by choosing suitable boundary distance functions in $\mathcal{R}(\mathcal{M})$ and employing them as local coordinates. This is a rather long procedure based on geometric properties of geodesics, hence is omitted. See Lemmas 2.14 and 3.32 of [104].

4.6   From $\mathcal{R}(\mathcal{M})$ to the Riemannian Metric

Let $H(x, \xi) = \frac{1}{2}g^{ij}(x)\xi_i\xi_j$, where $\left(g^{ij}\right) = \left(g_{ij}\right)^{-1}$. The equation of geodesics is written as Hamilton's canonical equation

$$\frac{dx^i}{dt} = \frac{\partial H}{\partial \xi_i}, \quad \frac{d\xi_i}{dt} = -\frac{\partial H}{\partial x^i}.$$

Let $x(0) = y$ and $\xi(0) = (\xi_{01}, \cdots, \xi_{0n})$ satisfy $g^{ij}(y)\xi_{0i}\xi_{0j} = 1$. Then, we have

$$g^{ij}(x(t))\xi_i(t)\xi_j(t) = 1.$$

Let $v(t) = (v^1(t), \cdots, v^n(t)) = \frac{dx(t)}{dt}$, and $v_0 = v(0)$. Then, $x(t)$ is a geodesic starting from y with initial direction v_0, which we denote as $x(t, y)$. Take an open set $U \subset S_y(\mathcal{M})$, where $S_y(\mathcal{M}) = \{v ; v \in T_y(\mathcal{M}), |v| = 1\}$, and assume that the map: $U \times (t_1, t_2) \ni (v_0, t) \to x(t, y)$ is a diffeomorphism. Let $t(x, y), \xi(x, y)$ be its inverse. It is well-known that $t(x, y) = d(x, y)$, moreover $t(x, y) = \int_y^x \xi_i dy^i$, hence $\dfrac{\partial t(x, y)}{\partial x^i} = \xi_i(x, y)$. We have, therefore,

$$\left(\mathrm{grad}_x t(x, y)\right)^i = g^{ij}(x)\frac{\partial t(x, y)}{\partial x^j} = v^i(x, y).$$

Now, we show that the Riemannian metric of $\mathcal{M}$ is reconstructed from BSD. We know already that BSD determines the boundary distance function. For $x_0 \in \mathcal{M}$, let $z_0 \in \partial\mathcal{M}$ be such that $d(x_0, z_0) = d(x_0, \partial\mathcal{M})$. Then, there is a small open cone of directions $C \subset S_{x_0}(\mathcal{M})$ such that the geodesic starting from x_0 with initial direction in C hits $\partial\mathcal{M}$ transversally in a neighborhood W_0 of z_0. Then, the directions of the shortest geodesic from $z_0 \in W_0$ to x_0 form the cone $-C$ in $S_{x_0}(\mathcal{M})$.

Let U be a small neighborhood of x_0. For $x \in U$ and $z \in W_0$, we consider $d(x, z)$. Passing to the Hamilton equation, we have $d(x, z) = t(x, z)$. We then have $g^{ij}(x_0)\xi_i(x_0, z)\xi_j(x_0, z) = 1$. We can compute $\xi(x_0, z)$ as $\xi_i(x_0, z) = \frac{\partial d}{\partial x^i}(x_0, z)$. Let z vary on W_0. Then, since $\xi(x_0, z)$ varies over an open set in $S_{x_0}^*(\mathcal{M}) = $ the unit sphere in the cotangent space $T_{x_0}^*(\mathcal{M})$, we can reconstruct the metric tensor $g^{ij}(x_0)$. We have thus proven the following theorem.

Theorem 4.2 *One can uniquely reconstruct the manifold $\mathcal{M}$ and the Riemannian metric $g_{ij}(x)$ from BSD.*

4.7 Wave Propagation

Let us return to Theorem 4.1. As will be seen below, it is related to the fundamental problems in wave propagation. First, we consider the domain of influence. For a subset $A \subset \mathcal{M}$, we put

$$D_+(A, t_0) = \{(x, t) ; x \in \mathcal{M}(A, t_0 \mp t), \ 0 \leq \bot t \leq t_0\},$$

$$D(A, t_0) = D_+(A, t_0) \cup D_-(A, t_0).$$

Theorem 4.3 *Take $t_0 > 0$ and a bounded open set $A \subset \mathcal{M}$ arbitrarily. Let u be a solution to IBVP*

$$\begin{cases} \partial_t^2 u = \Delta_g u \quad in \quad \mathcal{M} \times \mathbf{R}, \\ u = \partial_t u = 0, \quad on \quad \mathcal{M}(A, t_0) \quad at \quad t = 0, \\ \partial_\nu u = 0, \quad on \quad D(A, t_0) \cap (\partial\mathcal{M} \times \mathbf{R}). \end{cases}$$

Then, $u = 0$ in $D(A, t_0)$.

Compared with the one-dimensional wave equation of the string $\partial_t^2 u = c^2 \partial_x^2 u$, (g^{ij}) plays the role of the square of the velocity of the wave. Usually, the uniqueness theorem like Theorem 4.3 is proven by the energy inequality over a cone in (x, t)-space with slope corresponding to the maximal speed of propagation, which is then constant. However, Theorem 4.3 takes into account of variable speed of propagation, hence we need to consider more accurate propagation cone. A brief sketch of the proof is as follows. We consider $D_+(A, t_0)$.

(1) Split $D_+(A, t_0)$ into thin slices $D_+^{(k)}(A, t_0) = \{(x, t); x \in D_+(A, t_0), k\epsilon \leq t \leq (k+1)\epsilon\}$, $k = 0, 1, \cdots, [t_0/\epsilon]$.
(2) Take a cone $D_k(\epsilon)$ such that $D_k(\epsilon) \subset D_+^{(k)}(A, t_0)$, and $u(t) = 0$ in $D_k(\epsilon)$.
(3) Let $\epsilon \to 0$ so that $\cup_k D_k(\epsilon) \to D_+(A, t_0)$.

Of course, the delicate step is (2). The idea consists in constructing a family of small cones and freezing the metric to be constant on each cone, on which lateral side the normal $\nu = (\nu_1, \cdots, \nu_n, \nu_t)$ has the property $\nu_t \geq (g^{ij}\nu_i\nu_j)^{1/2}$. On this approximate cone, one can derive the energy inequality hence the uniqueness.[4]

For a solution u^f of IBVP 4.1, let $\mathrm{supp}_x u^f(t)$ be the support with respect to x of $u^f(x, t)$. Theorem 4.3 implies the following propagation property.

Lemma 4.7 *Let $S \subset \partial\mathcal{M}$ be an open set, and u^f the solution to IBVP (4.1). Assume that $\mathrm{supp}\, f \subset S \times (0, \infty)$. Then for any $t_0 > 0$*

$$\mathrm{supp}_x u^f(t_0) \subset \mathcal{M}(S, t_0).$$

---

[4]Let us consider the case in which $\mathcal{M}$ is a domain in $\mathbf{R}^n$. The general case can be proved in the same way by taking local coordinates. Take a large constant $C_0 > 0$ and put for $j = (j_1, \cdots, j_n) \in \mathbf{Z}^n$, $p_{j,\epsilon} = (j_1\epsilon/C_0, \cdots, j_n\epsilon/C_0)$. We extend $(g^{ij}(x))$ outside $\mathcal{M}$ and put $G^{(j,\epsilon)} = (g^{\alpha\beta}(p_{j,\epsilon})) + \epsilon C_0 I_n$. We also put $\mathcal{M}_\epsilon(A, t_0) = \{x \in \mathcal{M}(A, t_0); d(x, \mathcal{M}(A, t_0)) > \epsilon\}$. Take a finite set $J(\epsilon) = \{j; p_{j,\epsilon} \in \mathcal{M}_\epsilon(A, t_0)\}$. Letting $d_{j,\epsilon}$ be the distance defined by the metric $(G^{j,\epsilon})^{-1}$, we define $D(j, \epsilon) = \{(x, t); x \in \mathcal{M}, d_{j,\epsilon}(x, p_{j,\epsilon}) \leq \epsilon - t, 0 \leq t \leq \epsilon/C_0\}$, $D_1(\epsilon) = \cup_{j \in J(\epsilon)} D(j, \epsilon)$. One can then prove the energy inequality and show that $u(t) = 0$ in $D_1(\epsilon)$. The next step is to start from the time $t = \epsilon/C_0$ instead of $t = 0$, and $D_1(\epsilon)$ instead of $D(A, t_0)$. One can then construct $D_2(\epsilon)$ as above. Repeating this procedure, one can construct $D_1(\epsilon), \cdots, D_N(\epsilon)$. Putting $D^{(N)} = \cup_{i=1}^N D_i(\epsilon)$, we have $u = 0$ in $D^{(N)}$. We can then show that $D^{(N)} \to D_+(A, t_0)$. See Lemma 6.4.1 of [91] for details.

In fact, for $(y_0, t_0) \notin \mathcal{M}(S, t_0)$, there exists $r > 0$ such that

$$\{(x, t) \,; \, x \in \mathcal{M}, \, d(x, S) \leq t, \, 0 \leq t \leq t_0\} \cap D((B_r(y_0), t_0) = \emptyset,$$

where $B_r(y_0)$ is the ball of radius $r > 0$ centered at y_0. Theorem 4.3 then proves $u^f(y_0, t_0) = 0$.

Our next aim is to show that $\mathcal{M}(S, t_0)$ is filled by $\mathrm{supp}_x u^f(t_0)$ for a suitable choice of f. For any $t_0 > 0$, the *controllability operator* $\mathscr{C}_{t_0}^S$ is defined by

$$\mathscr{C}_{t_0}^S : L^2(S \times (0, t_0)) \ni f \to u^f\big|_{t=t_0} \in L^2(\mathcal{M}(S, t_0)).$$

The following fact is crucial for the BC-method.

Theorem 4.4 $\overline{\mathrm{Ran}\,(\mathscr{C}_{t_0}^S)} = L^2(\mathcal{M}(S, t_0))$.

To prove this theorem, we consider the adjoint problem. For the solution v^ψ of the following IBVP

$$\begin{cases} \partial_t^2 v = \Delta_g v & \text{in } \mathcal{M} \times \mathbf{R}, \\ v\big|_{t=t_0} = 0, \quad \partial_t v\big|_{t=t_0} = \psi \in L^2(\mathcal{M}(S, t_0)), \\ \partial_\nu v\big|_{\partial \mathcal{M} \times (0, t_0)} = 0, \end{cases}$$

The *observability operator* is defined by

$$\mathscr{O}_{t_0}^S : L^2(\mathcal{M}(S, t_0)) \ni \psi \to v^\psi\big|_{S \times (0, t_0)} \in L^2(S \times (0, t_0)).$$

By integration by parts, one has

$$(C_{t_0}^S f, \psi)_{L^2(\mathcal{M}(S, t_0))} = -(f, \mathscr{O}_{t_0}^S \psi)_{L^2(S \times (0, t_0))},$$

i.e., $\mathscr{C}_{t_0}^S = -(\mathscr{O}_{t_0}^S)^*$. Then, Theorem 4.4 follows from

Theorem 4.5 $\mathrm{Ker}\,\mathscr{O}_{t_0}^S = \{0\}$.

To prove this theorem, it is sufficient to show

Lemma 4.8 *Assume that v satisfies*

$$\begin{cases} \partial_t^2 v = \Delta_g v & \text{in } \mathcal{M} \times \mathbf{R}, \\ v\big|_{t=t_0} = 0, \quad \partial_t v\big|_{t=t_0} = \psi \in L^2(\mathcal{M}(S, t_0)), \\ \partial_\nu v\big|_{\partial \mathcal{M} \times (0, t_0)} = 0, \quad v\big|_{S \times (0, t_0)} = 0. \end{cases}$$

Then, $\partial_t v\big|_{t=t_0} = 0$.

This lemma in turn follows from the following theorem due to Tataru [185].

Theorem 4.6 *Let $S \subset \partial \mathcal{M}$ be an open set, and $u \in H^1_{loc}(\mathcal{M} \times (-t_0, t_0))$ be a solution to IBVP*

$$
\begin{cases}
\partial_t^2 u = \Delta_g u, & in \quad \mathcal{M} \times (-t_0, t_0), \\
\partial_\nu u \big|_{\partial \mathcal{M} \times (-t_0, t_0)} = 0, \quad u \big|_{S \times (-t_0, t_0)} = 0.
\end{cases}
$$

Then $u = 0$ in $D(S, t_0)$.

It is worth recalling here a brief history of the study of the uniqueness for PDE. Note that for a differential operator $P(x, D_x) = \sum_{|\alpha| \leq m} p_\alpha(x) D_x^\alpha$ defined on an open set U in $\mathbf{R}^n$, its principal symbol is defined by $P_m(x, \xi) = \sum_{|\alpha| = m} p_\alpha(x, \xi)$. A surface Σ of codimension 1 in U is said to be non-characteristic to $P(x, D_x)$ if $P_m(x, \nu_x) \neq 0$ for any $x \in \Sigma$ and normal ν_x to Σ at x. The first general theorem for the uniqueness of linear PDE is due to Holmgren [75].

Theorem 4.7 *Let u be a classical solution to the equation $P(x, D_x)u = 0$ with analytic coefficients. If $u = 0$ in one side of a non-characteristic surface Σ, then $\mathrm{supp}\, u \cap \Sigma = \emptyset$, i.e. $u = 0$ near Σ.*

This theorem was extended by John [98] for the analytic case, Robbiano [170] and Hörmander [77] for the C^∞ case, and finally Tataru arrived at the final result Theorem 4.6. The importance of non-analyticity should be strongly emphasized in the application to inverse problems. For time-independent operators $P(x, D_x)$, the problem is thus settled, however, it is still open for the time-dependent operators $P(t, x, D_x)$ except for the case P is analytic in t (see [6]). We should also mention that the Carleman estimate plays a key role here. For the proof of Theorem 4.6, see [104], p. 177.

4.8 Other Spectral Data

We have discussed the inverse problem from BSD. However, the BC-method works well also for the other spectral data. Consider the boundary value problem

$$
\begin{cases}
-\dfrac{1}{\sqrt{g}} \partial_i \left(\sqrt{g} g^{ij} \partial_j u \right) = zu & in \quad \mathcal{M}, \\
\partial_\nu u = f & on \quad \partial \mathcal{M}.
\end{cases}
\tag{4.3}
$$

Let $\sigma_p(-\Delta_g^{(N)})$ be the set of eigenvalues of Neumann Laplacian. It is well-known that for any $f \in H^{1/2}(\partial \mathcal{M})$ and $z \notin \sigma_p(-\Delta_g^{(N)})$, there exists a unique solution

$u(z) \in H^2(\mathcal{M})$ of (4.3), and $u(z)$ is an $H^2(\mathcal{M})$-valued analytic function of $z \in \mathbf{C} \setminus \sigma_p(-\Delta_g^{(N)})$. We define the *elliptic Neumann-to-Dirichlet map* $\Lambda_{ell}(z)$ by

$$\Lambda_{ell}(z)f = u(z)\big|_{\partial \mathcal{M}}.$$

Let $0 = \lambda_0 < \lambda_1 < \cdots$ be the eigenvalues without counting multiplicities, and $\varphi_{i,1}, \cdots, \varphi_{i,m_i}$ the orthonormal eigenvectors associated with λ_i, where m_i is the multiplicity of λ_i. By integration by parts, $\Lambda_{ell}(z)$ has a formal integral kernel written as

$$\Lambda_{ell}(x, y; z) = \sum_{i=1}^{\infty} \frac{\sum_{j=1}^{m_i} \varphi_{i,j}(x)\varphi_{i,j}(y)}{\lambda_i - z}, \quad x, y \in \partial \mathcal{M}. \tag{4.4}$$

This shows that the BSD and the elliptic N-D map are equivalent. It can be made precise as follows. Let $\Gamma = \partial \mathcal{M}$, $r_\Gamma \in \mathbf{B}(H^1(\mathcal{M}); H^{1/2}(\Gamma))$ be the trace operator, and $\delta_\Gamma = r_\Gamma^* \in \mathbf{B}(H^{-1/2}(\Gamma); H^1(\mathcal{M})^*)$ its adjoint

$$(\delta_\Gamma f, w)_{L^2(\mathcal{M})} = (f, r_\Gamma w)_{L^2(\partial \mathcal{M})}.$$

Then, we have

$$\Lambda_{ell}(z) = \delta_\Gamma^* R(z)\delta_\Gamma, \tag{4.5}$$

where $R(z) = (-\Delta_g^{(N)} - z)^{-1}$. The right-hand side of (4.5) is a $\mathbf{B}(H^{-1/2}(\Gamma); H^{1/2}(\Gamma))$-valued meromorphic function of $z \in \mathbf{C}$ with poles at $\lambda_i \in \sigma(-\Delta_g^{(N)})$ whose residue is the restriction of the associated eigenprojection to $\partial \mathcal{M}$. This justifies (4.4). Note that $\sum_{j=1}^{m_i} \varphi_{i,j}(x)\varphi_{i,j}(y)\big|_{\Gamma \times \Gamma}$ is independent of the choice of orthonormal eigenvectors. Therefore, what we really use in the BC-method is

$$\left\{ \left(\lambda_i, \sum_{j=1}^{m_i} \varphi_{i,j}(x)\varphi_{i,j}(y)\big|_{\Gamma \times \Gamma}\right); i = 0, 1, 2, \cdots \right\},$$

which should be called the *boundary spectral projection*.

Let $u^f(t)$ be the solution to (4.1). We define the *hyperbolic Neumann-to-Dirichlet map* Λ_{hyp} by

$$\Lambda_{hyp} f(t) = u^f(t)\big|_{\partial \mathcal{M}}.$$

The following equality is called *Blagovestchenskii's identity*.

Lemma 4.9 *Let $T > 0$ and $f, h \in C_0^\infty(\partial \mathcal{M} \times (0, T])$. Then,*

$$(u^f(T), u^h(T))_{L^2(\mathcal{M})}$$

$$= \frac{1}{2} \int_D \left\{ (\partial_\nu f(t), \Lambda_{hyp} h(s))_{L^2(\partial \mathcal{M})} - (\Lambda_{hyp} f(t), \partial_\nu u^h(s))_{L^2(\partial \mathcal{M})} \right\} ds dt,$$

where D is defined by $D = \{(t, s)\,;\, 0 \le t \le T,\ t \le s \le 2T - t\}$.[5]

In view of Corollary 4.1, comparing Lemma 4.9 with Lemma 4.1, we see that the hyperbolic N-D map also serves as a spectral data for the BC-method.

We consider the relation between $\Lambda_{ell}(z)$ and $\Lambda_{hyp}(t)$. Take non-trivial $\psi(t) \in C_0^\infty((0, \infty))$ and let $u(t)$ be the solution of (4.1) with f replaced by $\psi(t)f$, where $f \in C^\infty(\partial \mathcal{M})$. For $z \in \mathbf{C}_+$, we put

$$v(z) = \int_{-\infty}^\infty e^{it\sqrt{z}} u(t) dt.$$

Since $u(t) = 0$ for $t < 0$, the integration in t is actually performed for $t \ge 0$, and $v(z)$ is analytic for $z \in \mathbf{C}_+$. Then $v(z)$ satisfies (4.3), where f is replaced by $\phi(z)f$ with

$$\phi(z) = \int_{-\infty}^\infty e^{it\sqrt{z}} \phi(t) dt.$$

Let $w(z) = v(z)/\phi(z)$. Then, $w(z)$ is meromorphic on $\mathbf{C}_+$, continuous on $\overline{\mathbf{C}_+}$ except for a discrete set, and for $\lambda \in \mathbf{R}$, $w(\lambda) = f$ on $\partial \mathcal{M}$. Therefore $w(\lambda)|_{\partial \mathcal{M}}$ coincides with $\Lambda_{ell}(z)f$, which shows that $\Lambda_{hyp}(t)$ determines $\Lambda_{ell}(z)$.

Conversely, $\Lambda_{ell}(z)$ determines $\Lambda_{hyp}(t)$. In fact, let $u(t)$ be the solution of (4.1) and put for $z \in \mathbf{C}_+ v(z) = \int_{-\infty}^\infty e^{it\sqrt{z}} u(t) dt$. Then, $v(z)|_{\partial \mathcal{M}} = \Lambda_{ell}(z)h(z)$, where $h(z) = \int_{-\infty}^\infty e^{it\sqrt{z}} f(t) dt$. Passing to the inverse Fourier transform, we see that $\Lambda_{hyp}(t)$ is determined by $\Lambda_{ell}(z)$.

The *parabolic Neumann-to-Dirichlet map* map is defined by

$$\Lambda_{par}(t)f = u(t)|_{\partial \mathcal{M}},$$

---

[5]Let $w(t, s) = (u^f(t), u^h(s))_{\mathcal{M}}$. Then, we have $(\partial_t^2 - \partial_s^2)w(t, s) = (\partial_\nu u^f(t), u^h(s))_{\partial \mathcal{M}} - (u^f(t), \partial_\nu u^h(s))_{\partial \mathcal{M}}$, and $w|_{t=0} = \partial_t w(t)|_{t=0} = w|_{s=0} = \partial_s w|_{s=0} = 0$. Solving this wave equation, we obtain the lemma.

where $u(t)$ is a solution to

$$
\begin{cases}
\partial_t u = \dfrac{1}{\sqrt{g}}\partial_i\left(\sqrt{g}\,g^{ij}\partial_j u\right), & t > 0, \\[2mm]
u\big|_{t=0} = 0, \\[2mm]
\partial_\nu u(t)\big|_{\partial\mathcal{M}} = f.
\end{cases}
$$

One can prove that $\Lambda_{par}(t)$, $\Lambda_{hyp}(t)$, $\Lambda_{ell}(z)$, and BSD determine each other.

4.9 Inverse Scattering

Let us briefly look at the recent results on the spectral theory and inverse scattering on non-compact manifolds. Melrose [143] proposed a framework dealing with the scattering theory on non-compact manifolds as a boundary value problem with singular metric at the boundary. Utilizing this framework, Sá Barreto [174] showed that asymptotically hyperbolic manifolds are reconstructed from the S-matrix. See also [71]. The hyperbolic manifold is an interesting example in that it is realized by the action of discrete groups on the hyperbolic space $\mathbf{H}^n$. In general, its ends have cusps and its metric has conical singularities, i.e. it is an *orbifold*, and appears in number theory. In [96], both of cusp and orbifold structure are treated for the case $n = 2$. In particular, the *generalized S matrix* is introduced, and it is shown that the knowledge of generalized S-matrix determines the Fuchsian group. For higher dimensions, see [94]. Spectral properties for non-compact manifolds depend heavily on its volume growth. The slower the volume, the harder the analysis. In particular, the asymptotically cylindrical manifolds, i.e. the case in which the metric behaves like $ds^2 \sim (dr)^2 + (dx)^2$ in (4.7), have exceptional features. However, one can still prove that the S-matrix determines the manifold (see [93]). This suggests us a possibility of unified result for inverse scattering on non-compact manifolds.

To study the inverse scattering on non-compact manifolds, the first issue is the class of manifolds as large and tractable as possible. Let us begin with a simple case. Let $\mathcal{M}$ be an n-dimensional, connected, non-compact Riemannian manifold. Fix a point $p_0 \in \mathcal{M}$ arbitrarily and let $\mathcal{M}_1 = \{p \in \mathcal{M}\,;\,r = \text{dist}(p, p_0) > 1\}$, where $\text{dist}(p, p_0)$ is the geodesic distance from p_0 to p. Assume that $\mathcal{M}_1$, henceforth called the *end* of $\mathcal{M}$, is diffeomorphic to $(1, \infty) \times M$, where M is an $(n - 1)$-dimensional compact manifold. Spectral properties of the Laplacian on $\mathcal{M}$ depend largely on the volume growth of the manifold at infinity. Letting $\sum_{i,j=1}^{n} g_{ij}(X)dX^i dX^j$ be the Riemannian metric on $\mathcal{M}$, and $g = \det(g_{ij})$, we assume that

$$
\frac{g'}{4g} = \frac{(n-1)c_0}{2} + c_1 r^{-\alpha} + \cdots, \qquad r \to \infty, \tag{4.6}
$$

where $' = \frac{\partial}{\partial r}$, $c_0, c_1, \alpha \in \mathbf{R}$ and $\alpha > 0$ is a constant. Therefore, $\log g$ behaves at most linearly as $r \to \infty$. This is a natural assumption, since outside this region, the Laplacian $-\Delta_g$ may not have continuous spectrum. When (4.6) holds, the essential spectrum of $-\Delta_g$ is the interval $[E_0, \infty)$, where $E_0 = ((n-1)c_0/2)^2$. To study more detailed spectral properties, we assume that the metric on $\mathcal{M}_1$ has the form

$$ds^2 = (dr)^2 + \rho(r)^2 h(r, x, dx), \tag{4.7}$$

where $h(r, x, dx) = \sum_{i,j=1}^{n-1} h_{ij}(r, x) dx^i dx^j$ is a metric on M depending smoothly on $r > 0$, and

$$h(r, x, dx) \to h_M(x, dx), \quad \text{as } r \to \infty,$$

$h_M(x, dx)$ being positive definite on M. If $\rho(r) \to \infty$ as $r \to \infty$, $\mathcal{M}_1$ is growing at infinity, and if $\rho(r) \to 0$, $\mathcal{M}_1$ is shrinking. In the former case, $\mathcal{M}_1$ is said to be a *regular end*, and in the latter case, a *cusp end*.

With these preliminaries in mind, let us consider a connected, non-compact Riemannian manifold $\mathcal{M}$ of the following form

$$\mathcal{M} = \mathcal{K} \cup \mathcal{M}_1 \cup \cdots \cup \mathcal{M}_{N+N'}, \tag{4.8}$$

where $\mathcal{K}$ is a relatively compact open set and $\mathcal{M}_j$, $j = 1, \cdots, N + N'$, is an unbounded open set such that $\mathcal{M}_i \cap \mathcal{M}_j = \emptyset$ if $i \neq j$. Assume that $\mathcal{M}_j$ is diffeomorphic to $(1, \infty) \times M_j$, where M_j is compact and $\dim M_j = n - 1$, $\mathcal{M}_j$ is a regular end for $1 \leq j \leq N$, and a cusp end for $N + 1 \leq j \leq N + N'$. We are interested in the case in which

$$\text{either } \rho_j(r) \sim e^{c_j r}, \text{ or } \rho_j(r) \sim r^{\beta_j},$$

where $c_j > 0$, $\beta_j > 0$ on regular ends, and $c_j < 0$, $\beta_j < 0$ on cusp ends. If $\rho_j(r) \sim r$, $\mathcal{M}_j$ is asymptotically Euclidean, and if $\rho_j(r) \sim e^{c_j r}$, $\mathcal{M}_j$ is asymptotically hyperbolic. For a real constant κ, let S^κ be the set of smooth functions on $\mathcal{M}$ having the following property on each end $\mathcal{M}_j$:

$$S^\kappa \ni f \iff |\partial_r^\ell \partial_x^\alpha f(r, x)| \leq C_{\ell \alpha} (1 + r)^{\kappa - \ell}, \quad \forall \ell, \alpha.$$

Assume that for $j = 1, \cdots, N + N'$,

$$\frac{\rho_j'(r)}{\rho_j(r)} - c_{0,j} \in S^{-\alpha_{0,j}}, \quad h_j(r, x, dx) - h_{M_j}(x, dx) \in S^{-\gamma_{0,j}},$$

where $\alpha_{0,j} > 0$, $\gamma_{0,j} > 1$. Moreover, for regular ends $\mathcal{M}_j$, $j = 1, \cdots, N$,

$$\frac{\rho_j'(r)}{\rho_j(r)} \geq \frac{\beta_{0,j}}{r},$$

where $\beta_{0,j} > 1/2$.[6] Let $I_0 = (0, 1]$, $I_\ell = (2^{\ell-1}, 2^\ell)$, $\ell \geq 1$, and define $\mathscr{B}$ to be the set of L^2-functions on $\mathscr{M}$ such that

$$\|f\|_{\mathscr{B}} = \|f\|_{L^2(\mathscr{K})} + \sum_{j=1}^{N+N'} \|f\|_{\mathscr{B}(\mathscr{M}_j)} < \infty,$$

$$\|f\|_{\mathscr{B}(\mathscr{M}_j)} = \sum_{\ell=0}^{\infty} 2^{\ell/2} \|f\|_{L^2(I_\ell, L^2(M_j), \rho^{n-1}(r)dr)}$$

on each end. Its dual space is identified with the set of L^2_{loc}-functions on $\mathscr{M}$ such that

$$\|u\|_{\mathscr{B}^*} = \|u\|_{L^2(\mathscr{K})} + \sum_{j=1}^{N+N'} \|u\|_{\mathscr{B}^*(\mathscr{M}_j)} < \infty,$$

$$\|u\|_{\mathscr{B}^*(\mathscr{M}_j)} = \left(\sup_{R>1} \frac{1}{R} \int_0^R \|u(r)\|^2_{L^2(M_j)} \rho^{n-1}(r)dr \right)^{1/2}$$

on each end. We put $E_{0,j} = ((n-1)c_{0,j}/2)^2$ and

$$E_0 = \min_{1 \leq j \leq N+N'} E_{0,j}, \quad E_{0,reg} = \min_{1 \leq j \leq N} E_{0,j}.$$

Let $H = -\Delta_g$ and put $R(z) = (H - z)^{-1}$, also $\mathscr{T} = \{E_{0,1}, \cdots, E_{0,N+N'}\}$.

We have given a typical solution to the forward problem in Sect. 3.2 for the case of $\mathbf{R}^n$. It is extended to $\mathscr{M}$ in the same way. The limiting absorption principle is stated as follows.

Theorem 4.8

(1) $\sigma_e(H) = [E_0, \infty)$, *and* H *has no eigenvalues in* $(E_{0,reg}, \infty)$. *The eigenvalues in* $(E_0, E_{0,reg}) \setminus \mathscr{T}$ *are discrete and may accumulate only in* $\mathscr{T}$.

(2) *Let* $\mathscr{E} = \sigma_p(H) \cup \mathscr{T}$. *Then, for any* $\lambda \in (E_0, \infty) \setminus \mathscr{E}$, *there exists a weak ∗-limit* $\lim_{\epsilon \to 0} R(\lambda \pm i\epsilon) = R(\lambda \pm i0)$, *i.e.*

$$(R(\lambda \pm i0)f, g) = \lim_{\epsilon \to 0} (R(\lambda \pm i\epsilon)f, g), \quad f, g \in \mathscr{B}.$$

---

[6]The condition for $\beta_{0,j}$ is weakened to be $\beta_{0,j} > 0$. However, it makes the following arguments more complicated. We also need a little more assumptions for cusp ends, which are omitted for the sake of simplicity.

(3) For any compact set $I \subset (E_0, \infty) \setminus \mathcal{E}$, there exists a constant $C > 0$ such that

$$\|R(\lambda \pm i0)f\|_{\mathscr{B}^*} \leq C\|f\|_{\mathscr{B}}, \quad f \in \mathscr{B}$$

We derive an asymptotic expansion of the resolvent. Define

$$f \in \mathscr{B}_0^*(\mathscr{M}_j) \iff \lim_{R \to \infty} \frac{1}{R} \int_0^R \|f(r)\|_{L^2(M_j)} \rho_j(r)^{(n-1)/2} dr = 0,$$

$$f \simeq g \quad \text{in} \quad \mathscr{M}_j \iff f - g \in \mathscr{B}_0^*(\mathscr{M}_j).$$

Lemma 4.10 *Let*

$$\phi_j(r, \lambda) = \sqrt{\lambda - \frac{(n-1)^2}{4}\left(\frac{\rho_j'(r)}{\rho_j(r)}\right)^2}, \quad \Phi_j(r, \lambda) = \int_0^r \phi_j(t, \lambda)dt,$$

$$C_j(\lambda) = \left(\frac{\pi}{\sqrt{\lambda - E_{0,j}}}\right)^{1/2}.$$

(1) For $1 \leq j \leq N$, $f \in \mathscr{B}$ and $\lambda \in (E_{0,j}, \infty) \setminus \mathcal{E}$, there exists a unique $a_j(\lambda, x) \in L^2(M_j)$ such that

$$R(\lambda \pm i0)f \simeq C_j(\lambda)\rho_j(r)^{-(n-1)/2}e^{\pm i\Phi_j(r,\lambda)}a_j(\lambda, x), \quad on \quad \mathscr{M}_j. \quad (4.9)$$

(2) For $N+1 \leq j \leq N+N'$, $f \in \mathscr{B}$ and $\lambda \in (E_{0,j}, \infty) \setminus \mathcal{T}$, there exists a unique $a_j(\lambda) \in \mathbf{C}$ such that

$$R(\lambda \pm i0)f \simeq C_j(\lambda)\rho_j(r)^{-(n-1)/2}e^{\pm i\Phi_j(r,\lambda)}a_j(\lambda) \quad on \quad \mathscr{M}_j.^{[7]} \quad (4.10)$$

We construct a spectral representation of $-\Delta_g$. Define for $\lambda > E_{0,j}$,

$$\mathscr{F}_j^{(\pm)}(\lambda)f = \begin{cases} a_j(\lambda, x), & \text{for} \quad 1 \leq j \leq N, \\ a_j(\lambda), & \text{for} \quad N+1 \leq j \leq N+N', \end{cases}$$

and $\mathscr{F}_j^{(\pm)}(\lambda) = 0$ for $\lambda < E_{0,j}$. We also put

$$\mathbf{h}_\infty = \bigoplus_{j=1}^{N+N'} \mathbf{h}_{\infty,j}, \quad \mathbf{h}_{\infty,j} = \begin{cases} L^2(M_j) & \text{for} \quad 1 \leq j \leq N, \\ \mathbf{C} & \text{for} \quad N+1 \leq j \leq N+N', \end{cases}$$

$$\widehat{\mathscr{H}} = \bigoplus_{j=1}^{N+N'} L^2((E_{0,j}, \infty), \mathbf{h}_{\infty,j}; d\lambda).$$

---

[7]For the sake of simplicity, we assume here that the volume of M_j is equal to 1.

Then, for any $\lambda \in (E_0, \infty) \setminus \mathcal{E}$, $\mathscr{F}^{(\pm)}(\lambda) \in \mathbf{B}(\mathscr{B}; \mathbf{h}_\infty)$. Moreover, we have

$$\frac{1}{2\pi i}(R(\lambda + i0)f - R(\lambda - i0)f, g) = (\mathscr{F}^{(\pm)}(\lambda)f, \mathscr{F}^{(\pm)}(\lambda)g)_{\mathbf{h}_\infty}.$$

Theorem 4.9

(1) Define $(\mathscr{F}^{(\pm)}f)(\lambda) = \mathscr{F}^{(\pm)}(\lambda)f$ for $f \in \mathscr{B}$. Then, $\mathscr{F}^{(\pm)}$ is uniquely extended to a partial isometry with initial set $\mathscr{H}_{ac}(H)$ and final set $\widehat{\mathscr{H}}$.
(2) For any $f \in D(H)$, we have $(\mathscr{F}^{(\pm)}Hf)(\lambda) = \lambda(\mathscr{F}^{(\pm)}f)(\lambda)$ a. e. λ.
(3) $\mathscr{F}^{(\pm)}(\lambda)^* \in \mathbf{B}(\mathbf{h}_\infty; \mathscr{B}^*)$ is an eigenoperator of H in the sense that

$$(-\Delta_g - \lambda)\mathscr{F}^{(\pm)}(\lambda)^* = 0.$$

(4) Let $I_1 \subset I_2 \subset \cdots$ be a set of intervals such that $I_n \to (E_0, \infty)$. Then,

$$f = \lim_{n \to \infty} \int_{I_n} \mathscr{F}^{(\pm)}(\lambda)^*(\mathscr{F}^{(\pm)}f)(\lambda)d\lambda, \quad f \in \mathscr{H}_{ac}(H)$$

holds in the sense of strong limit in $L^2(\mathscr{M})$.

Let $c_j(\lambda)$ be the characteristic function of $(E_{0,j}, \infty)$ and put

$$\mathbf{h}_\infty(\lambda) = \bigoplus_{j=1}^{N+N'} c_j(\lambda)\mathbf{h}_{\infty,j}.$$

The eigenoperator $\mathscr{F}^{(\pm)}(\lambda)^* \in \mathbf{B}(\mathbf{h}_\infty; \mathscr{B}^*)$ characterizes the solution space for the Helmholtz equation in the following sense.

Theorem 4.10 *For $\lambda \in o_e(H) \setminus \mathcal{E}$, we have*

$$\{u \in \mathscr{B}^* ; (-\Delta_g - \lambda)u = 0\} = \mathscr{F}^{(\pm)}(\lambda)^*\mathbf{h}(\lambda).$$

We define the S-matrix $S(\lambda)$.

Theorem 4.11 *Let $\lambda \in (E_0, \infty) \setminus \mathcal{E}$. Then, for any $a^{(in)} \in \mathbf{h}_\infty(\lambda)$, there exist a unique $a^{(out)} \in \mathbf{h}_\infty(\lambda)$ and $u \in \mathscr{B}^*$ such that $(-\Delta_g - \lambda)u = 0$ in $\mathscr{M}$, moreover*

$$u \simeq C_j(\lambda)\rho_j(r)^{-(n-1)/2}\left(e^{-i\Phi_j(r,\lambda)}a^{(in)} - e^{i\Phi_j(r,\lambda)}a^{(out)}\right)$$

on each $\mathscr{M}_j$, $1 \leq j \leq N + N'$. The operator

$$S(\lambda) : \mathbf{h}_\infty(\lambda) \ni a^{(in)} \to a^{(out)} \in \mathbf{h}_\infty(\lambda)$$

is unitary.

We are now in a position to discussing the inverse scattering on $\mathcal{M}$. Our strategy is to split $\mathcal{M}$ into two parts $\mathcal{M} = \mathcal{M}_{ext} \cup \mathcal{M}_{int}$, where $\mathcal{M}_{ext} \cap \mathcal{M}_{int}$ is a smooth manifold of codimension 1. Assuming that we know the *exterior* domain $\mathcal{M}_{ext}$, we can obtain the N-D map for the *interior* domain $\mathcal{M}_{int}$ from the knowledge of the S-matrix. We can then apply the BC-method to $\mathcal{M}_{int}$. In view of (4.8), as $\mathcal{M}_{ext}$, we take either $\mathcal{M}_1$ (regular end) or $\mathcal{M}_{N+N'}$ (cusp end).

First we consider the inverse scattering from regular end. Assuming that $\mathcal{M}_1$ is a regular end, we put

$$\mathcal{M}_{ext} = \mathcal{M}_1 \cap \{r \geq 2\},$$

$$\mathcal{N} = \mathcal{M}_{int} = (\mathcal{M}_1 \cap \{r \leq 2\}) \cup \mathcal{K} \cup \mathcal{M}_2 \cup \cdots \cup \mathcal{M}_{N+N'}.$$

Then, $\mathcal{N}$ is a manifold with boundary $\partial \mathcal{N} = \mathcal{M}_1 \cap \{r = 2\}$, which is non-compact if $N + N' \geq 2$. We define the *interior* N-D map $\Lambda(\lambda)$ by

$$\Lambda(\lambda)f = u\big|_{\partial \mathcal{N}},$$

where u is a solution to the Neumann problem

$$\begin{cases} (-\Delta_g' - \lambda)u = 0 & \text{in} \quad \mathcal{N}, \\ \partial_\nu u = f & \text{on} \quad \partial \mathcal{N}, \end{cases}$$

Δ_g' being the Laplace operator of $\mathcal{N}$. Let $H' = -\Delta_g'$ in $\mathcal{N}$ with Neumann boundary condition on $\partial \mathcal{N}$. If $\mathcal{N}$ is unbounded, we need to assume that u satisfies the radiation condition to guarantee the uniqueness of the solution.[8] The S-matrix $S(\lambda)$ is an $(N + N') \times (N + N')$ matrix operator. We denote its $(1, 1)$ entry by $S_{11}(\lambda)$. The following lemma reduces the inverse scattering problem to the inverse boundary value problem.

Lemma 4.11 *Let* $\lambda \in (E_{0,1}, \infty) \setminus \mathscr{E}$.[9] *Then,* $S_{11}(\lambda)$ *and* $\Lambda(\lambda)$ *determine each other.*

Now suppose we are given two manifolds

$$\mathcal{M}^{(i)} = \mathcal{K}^{(i)} \cup \mathcal{M}_1^{(i)} \cup \cdots \cup \mathcal{M}_{N_i + N_i'}, \quad i = 1, 2,$$

having the above-mentioned properties. Note that the number of ends of $\mathcal{M}^{(1)}$ and $\mathcal{M}^{(2)}$ are not assumed to be equal. If $(1, 1)$ entries of the associated S-matrices coincide, due to Lemma 4.11, we can obtain the same N-D map $\Lambda(\lambda)$ on $\mathcal{N}^{(1)}$ and $\mathcal{N}^{(2)}$ from $S_{11}^{(1)}(\lambda) = S_{11}^{(2)}(\lambda)$. One can then apply the BC-method to $\mathcal{N}^{(i)}$ in the

---

[8]Namely, we assume that $(\partial_r + \frac{4g_j'}{g_j} - i\sqrt{\lambda - E_{0,j}})u \in \mathscr{B}_0^*(\mathcal{M}_j)$ for $2 \leq j \leq N + N'$.

[9]If $N + N' = 1$, we further need to assume that $\lambda \notin \sigma(H')$.

same way as above to conclude that $\mathcal{N}^{(1)}$ and $\mathcal{N}^{(2)}$ are isometric, although $\mathcal{N}^{(i)}$ may be non-compact. The inverse scattering for these manifolds is thus solved as follows.

Theorem 4.12 *Assume that $S_{11}^{(1)}(\lambda) = S_{11}^{(2)}(\lambda)$ for all $\lambda \in \sigma_e(H^{(1)}) \cap \sigma_e(H^{(2)})$. Assume further that $\mathcal{M}_1^{(1)}$ and $\mathcal{M}_1^{(2)}$ are isometric. Then, $\mathcal{M}^{(1)}$ and $\mathcal{M}^{(2)}$ are isometric.*

Next we consider the inverse scattering from cusp end. The components of the S-matrix for the cusp ends are complex numbers, i.e. $S_{jj}(\lambda) \in \mathbb{C}$ for $N + 1 \le j \le N + N'$. Therefore, they do not have enough information to determine the whole manifold $\mathcal{M}$. In fact, Zelditch [199] proved that the arithmetic surface is not determined from the S-matrix associated with the cusp end. To have more knowledge of the manifold, we need to enlarge the solution space of the Helmholtz equation $(-\Delta_g - \lambda)u = 0$. We pick up one cusp end $\mathcal{M}_j$, and assume without loss of generality that $j = N + N'$, moreover it is a pure cusp, i.e. its metric is of the form

$$ds^2 = (dr)^2 + \rho_j(r)^2 h_j(x, dx).$$

Then the Helmholtz equation $(-\Delta_{\mathcal{M}_j} - \lambda)u = 0$ in $\mathcal{M}_j$ can be solved by the separation of variables. Let Δ_{M_j} be the Laplace-Beltrami operator for M_j equipped with the metric $h_j(x, dx)$. Let $0 = \lambda_{0,j} < \lambda_{1,j} \le \lambda_{2,j} \le \cdots$ be the eigenvalues of $-\Delta_{M_j}$ with complete orthonormal system of eigenvectors $e_{\ell,j}(x)$. We put

$$\Phi_j(r, \lambda, B) = \int_0^r \sqrt{\frac{B}{\rho_j^2} - \lambda + \frac{(n^2 - 2n)}{4}\left(\frac{\rho_j'}{\rho_j}\right)^2 + \frac{(n-2)}{2}\left(\frac{\rho_j'}{\rho_j}\right)'}\, dr.$$

Then, there exist solutions $u_{\ell,j,\pm}$ of the equation

$$-u'' - \frac{(n-1)\rho_j'}{\rho_j}u' + \left(\frac{\lambda_{\ell,j}}{\rho_j^2} - \lambda\right)u = 0,$$

which behave like

$$u_{\ell,j,\pm} \sim \rho_j(r)^{-(n-1)/2}e^{\pm\Phi_j(r,\lambda_{\ell,j})}, \quad r \to \infty.$$

Let us introduce two spaces of sequences $\mathbf{A}_{j,\pm}$:

$$\mathbf{A}_{j,\pm} \ni \{c_{\ell,\pm}\}_{\ell=0}^{\infty} \iff \sum_{\ell=0}^{\infty} |c_{\ell,\pm}|^2 |u_{\ell,j,\pm}(r)|^2 < \infty, \quad \forall r > 0.$$

Letting $\chi_j(r) \in C^\infty(0, \infty)$ be such that on $\mathscr{M}_j$, $\chi_j(r) = 0$ for $r < 1$, $\chi_j(r) = 1$ for $r > 2$, we define the generalized incoming solution on $\mathscr{M}_j$ by

$$\Psi_j^{(in)} = \chi_j(r) \sum_{\ell=0}^{\infty} a_{\ell,j} u_{\ell,j,+}(r) e_{\ell,j}(x), \quad \{a_{\ell,j}\}_{\ell=0}^{\infty} \in \mathbf{A}_+,$$

which is (super) exponentially growing as $r \to \infty$, and the generalized outgoing solution on $\mathscr{M}_j$ by

$$\Psi_j^{(out)} = \chi_j(r) \sum_{\ell=0}^{\infty} b_{\ell,j} u_{\ell,j,-}(r) e_{\ell,j}(x), \quad \{b_{\ell,j}\}_{\ell=0}^{\infty} \in \mathbf{A}_-,$$

which is (super) exponentially decaying as $r \to \infty$. We also define the spaces of generalized scattering data by

$$\mathbf{h}_\infty^{(in)}(\lambda) = \left(\bigoplus_{j=1}^{N} c_j(\lambda) L^2(M_j) \right) \oplus \left(\bigoplus_{j=N+1}^{N+N'-1} c_j(\lambda) \mathbf{C} \right) \oplus \left(c_{N+N'}(\lambda) \mathbf{A}_{N+N',+} \right),$$

$$\mathbf{h}_\infty^{(out)}(\lambda) = \left(\bigoplus_{j=1}^{N} c_j(\lambda) L^2(M_j) \right) \oplus \left(\bigoplus_{j=N+1}^{N+N'-1} c_j(\lambda) \mathbf{C} \right) \oplus \left(c_{N+N'}(\lambda) \mathbf{A}_{N+N',-} \right).$$

One can then construct generalized scattering solutions to the Helmholtz equation on $\mathscr{M}$.

Theorem 4.13 *For any generalized incoming data $a^{(in)} \in \mathbf{h}_\infty^{(in)}(\lambda)$, there exist a unique solution u of the equation $(-\Delta_g - \lambda)u = 0$ and the outgoing data $a^{(out)} \in \mathbf{h}_\infty^{(out)}(\lambda)$ such that $u - \Psi_{N+N'}^{(in)} \in \mathscr{B}^*$, $u = \Psi_{N+N'}^{(in)} - \Psi_{N+N'}^{(out)}$ on $\mathscr{M}_{N+N'}$, and on the ends $\mathscr{M}_j$, $1 \le j \le N + N' - 1$, u has the asymptotic form as in (4.9) and (4.10).*

We call the operator

$$\mathscr{S}(\lambda) : \mathbf{h}_\infty^{(in)}(\lambda) \ni a^{(in)} \to a^{(out)} \in \mathbf{h}_\infty^{(out)}(\lambda)$$

the generalized S-matrix. It is an infinite matrix mapping $\{a_{\ell,N+N'}\}_{\ell=0}^{\infty}$ to $\{b_{\ell,N+N'}\}_{\ell=0}^{\infty}$, and the $(N + N', N + N')$-entry of the physical S-matrix $S(\lambda)$ is the $(0, 0)$-entry of the generalized S-matrix $\mathscr{S}(\lambda)$. Now, we split the manifold $\mathscr{M}$ into two parts $\mathscr{M} = \mathscr{M}_{ext} \cup \mathscr{M}_{int}$, where $\mathscr{M}_{ext} = \mathscr{M}_{N+N'} \cap \{r \ge 2\}$. Then, as in Lemma 4.11 the generalized S-matrix $\mathscr{S}(\lambda)$ determines the N-D map on $\mathscr{M}_{int}$. One can then apply the BC-method. The inverse scattering problem from cusp ends is now solved as follows.

Theorem 4.14 *Let $m^{(i)} = N^{(i)} + N'^{(i)}$. Assume that $\mathscr{S}_{m^{(1)}m^{(1)}}^{(1)}(\lambda) = \mathscr{S}_{m^{(2)}m^{(2)}}^{(2)}(\lambda)$ for all $\lambda \in \sigma_e(H^{(1)}) \cap \sigma_e(H^{(2)})$. Assume further that the cusp ends $\mathscr{M}_{m^{(1)}}^{(1)}$ and $\mathscr{M}_{m^{(2)}}^{(2)}$ are isometric. Then, $\mathscr{M}^{(1)}$ and $\mathscr{M}^{(2)}$ are isometric.*

Therefore, for the metric of the form (4.7) with the behavior

$$\rho(r) \sim e^{cr}, \quad \rho(r) \sim Cr^{\beta}, \quad r \to \infty, \tag{4.11}$$

where $C > 0$, $c, \beta \in \mathbf{R}$, we can reconstruct the metric from the knowledge of the (generalized) S-matrix for all energies.

The above results are given in [92], including the treatment of conical singularities. A brief outline was already given in [95].

Chapter 5
Other Topics

The range of application of scattering theory is widely spread. In this chapter, we pick up some topics which illustrate new aspects of inverse scattering. In Sect. 5.1, we show recent results on the formal determinacy of the mapping $V(x) \to S(k, \theta, \omega)$ for the wave equation with potential term, where the incident direction of waves is fixed. Section 5.2 deals with the inverse scattering for the Maxwell equation. We discuss the problem of determination of scalar dielectric permittivity and magnetic permeability, and the computation of Betti numbers for the case of anisotropic domain. In Sect. 5.3, we consider the inverse scattering on perturbed periodic lattices, discrete and quantum graphs. In Sect. 5.4, we mention a new aspect of the method of separation of variables in the inverse spectral problem.

5.1 Backscattering and Fixed Angle Scattering

The multi-dimensional inverse problem is usually overdetermined. For example, for the inverse problem of potential scattering associated with the Schrödinger operator $-\Delta + V(x)$ in $\mathbf{R}^n$, the spectral data is the scattering amplitude $A(\lambda, \theta, \omega)$ which depends on $2n - 1$ parameters while the object we seek is the potential $V(x)$ depending on n variables. One idea to make the problem to be properly (or formally) determined, i.e. the case in which the number of spectral parameters coincides with that of the unknown object, is to consider the *backscattering*, namely, to determine $V(x)$ from $A(\lambda; -\omega, \omega)$. In fact letting $V_{\lambda,\omega}(x) = V(x)e^{i\sqrt{\lambda}\omega \cdot x}$, by (3.4), we have, up to a constant depending on λ,

$$A(\lambda; \theta, \omega) = \widehat{V}(\sqrt{\lambda}(\theta - \omega)) - \widehat{U}_{\lambda,\omega}(\sqrt{\lambda}\theta), \qquad (5.1)$$

$$\widehat{U}_{\lambda,\omega}(\sqrt{\lambda}\theta) = (2\pi)^{-n/2} \int_{\mathbf{R}^n} e^{-i\sqrt{\lambda}\theta \cdot x} V(x)\big(R(\lambda + i0)V_{\lambda,\omega}\big)(x)dx.$$

© The Author(s), under exclusive licence to Springer Nature Singapore Pte Ltd. 2020
H. Isozaki, *Inverse Spectral and Scattering Theory*, SpringerBriefs
in Mathematical Physics 38, https://doi.org/10.1007/978-981-15-8199-1_5

Letting $\lambda = |\xi|^2/4$, $\theta = -\omega = \xi/|\xi|$, the first term of the right-hand side of (5.1) gives the Fourier transform of $V(x)$. It is then natural to ask the invertibility of the map

$$V(x) \to \widehat{A}(|\xi|/2; -\widehat{\xi}, \widehat{\xi}), \quad \widehat{\xi} = \xi/|\xi|. \tag{5.2}$$

Formal arguments were given by Moses [145] and Prosser [161]. Eskin and Ralston [56] proved that the map (5.2) is a local diffeomorphism on a certain open set of a Banach space of real potentials. Since this open set contains a small neighborhood of 0, small potentials are recovered from the backscattering. In general, it is hard to know global properties of the map (5.2). However, due to the regularity of the resolvent, the singularities of $V(x)$ is expected to be determined by $A(|\xi|/2, -\widehat{\xi}, \widehat{\xi})$.

It is also meaningful to study the backscattering for the wave equation

$$\begin{cases} (\partial_t^2 - \Delta + V(x))u(t,x) = 0, \\ u(t,x) = \delta(t - x \cdot \omega), \quad t << 0, \end{cases} \tag{5.3}$$

where $\omega \in S^{n-1}$. Assume that $n = 3$ and $V(x)$ is compactly supported. Then, for any $\theta \in S^2$ and $s \in \mathbf{R}$ the limit

$$\lim_{r \to \infty} ru(r\theta, r - s, \omega) = \frac{1}{2\pi} \int_{x \cdot \theta = \tau} u_t(x, \tau - s, \omega) dS_x =: B(s; \theta, \omega).$$

exists for any τ such that $\operatorname{supp} q(x) \subset \{x \cdot \theta < \tau\}$. The right-hand side, the time-dependent scattering amplitude, is often called the *far-field pattern*. Letting $u(t, x, \theta) = \delta(t - x \cdot \omega) + v(t, x)$, and $w(x) = \int_{-\infty}^{\infty} e^{ikt} v(t, x) dt$, we see that $w(x)$ is the solution of the Schrödinger equation

$$(-\Delta + V(x) - k^2)w = -V(x)e^{ik\omega \cdot x}$$

satisfying the radiation condition. As can be inferred from this fact, the scattering amplitude $A(\lambda; \theta, \omega)$ and the far-field pattern determine $B(s; \theta, \omega)$ each other. Slightly generalizing the question, we can also ask the problem of fixed angle inverse scattering, i.e. the inversion of the map

$$V(x) \to A_\omega(\lambda, \theta) = A(\lambda; \theta, \omega), \text{ or } B_\omega(s, \theta) = B(s; \theta, \omega)$$

with fixed incoming direction ω. Lots of works have been devoted to this backscattering problem, e.g. [20, 68, 144, 155, 169, 173, 179, 188]. Rakesh–Salo [162, 163] proved that the potential is uniquely determined from $B_\omega(s, \theta)$ or $A_\omega(\lambda, \theta)$ if $V(x)$ is symmetric with respect to the plane orthogonal to ω, i.e., letting $\omega = (0, \cdots, 0, 1)$, $V(x', x_n) = V(x', -x_n)$. The Carleman estimate plays a key role here.

5.2 Maxwell Equation

Letting $E(t, x)$ be the electric field strength and $B(t, x)$ the magnetic flux density, the Maxwell equation is

$$\begin{cases} \epsilon(x)\partial_t E = \text{curl } H, \\ \mu(x)\partial_t H = -\text{curl } E, \end{cases}$$

where $\epsilon(x)$, $\mu(x)$ are 3×3 positive definite matrices, called dielectric permittivity and magnetic permeability. For the boundary value problem in a domain $\Omega \subset \mathbf{R}^3$, the boundary condition

$$v \times E = 0 \quad \text{on} \quad \partial\Omega \tag{5.4}$$

is imposed, where v is the unit outer normal to the boundary. Let

$$H(\text{curl}; \Omega) = \{u \in L^2(\Omega)^3 ; \text{ curl } u \in L^2(\Omega)^3\},$$

and $H_0(\text{curl}; \Omega)$ be the closure of $C_0^\infty(\Omega)^3$ in $H(\text{curl}; \Omega)$. Let $\mathbf{H} = L^2(\Omega)^6$ be the Hilbert space equipped with the inner product $(f, g) = (\epsilon f_E, g_E) + (\mu f_H, g_H)$, where $f = (f_E, f_H), g = (g_E, g_H) \in \mathbf{H}$ and $(\ ,\ )$ in the right-hand side is the standard inner product of $L^2(\Omega)^3$. Letting $\mathcal{M} = \begin{pmatrix} \epsilon(x) & 0 \\ 0 & \mu(x) \end{pmatrix}$, $\mathcal{L} = i \begin{pmatrix} 0 & \text{curl} \\ \text{curl} & 0 \end{pmatrix}$, the operator $\mathcal{H} = \mathcal{M}(x)^{-1}\mathcal{L}$ with domain $D(\mathcal{H}) = H_0(\text{curl}; \Omega) \times H(\text{curl}; \Omega)$ is self-adjoint in $\mathbf{H}$. Let $N(\mathcal{H}) = \{f \in D(\mathcal{H}); \mathcal{H}f = 0\}$, $\mathcal{P}_0$ be the projection to $N(\mathcal{H})$ and $\mathcal{P} = 1 - \mathcal{P}_0$. Let $\chi_R(x)$ be the characteristic function of $\Omega \cap \{|x| < R\}$. Then, by using the result of Weber [193], one can show that $\chi_R(\mathcal{H} - z)^{-1}$ is compact on $\mathbf{H}$ for any $R > 0$.[1] Assuming that $|\epsilon(x) - \epsilon_0 I_3| + |\mu(x) - \mu_0 I_3| \leq C(1 + |x|)^{-\delta}$ for some $\delta > 1$, where ϵ_0, μ_0 are positive constants, and letting $\mathcal{H}_0$ be $\mathcal{H}$ with $\epsilon(x), \mu(x)$ replaced by ϵ_0, μ_0, one can show the existence and completeness of wave operators $W_\pm = \text{s} - \lim_{t \to \pm\infty} e^{it\mathcal{H}} e^{-it\mathcal{H}_0} P_{ac}(\mathcal{H}_0)$, where $P_{ac}(\mathcal{H}_0)$ is the projection onto the absolutely continuous subspace of $\mathcal{H}_0$. Then, the S-matrix $S(\lambda)$, $\lambda \in \mathbf{R} \setminus \{0\}$, is defined as a unitary operator on $L^2(S^2)^6$ in the same way as in the case of Schrödinger operators.

The reconstruction problem of scalar $\epsilon(x), \mu(x)$ is solved by Ola, Päivärinta, and Sommersalo [154] in the form of inverse boundary value problem. For a two-dimensional surface $S \subset \mathbf{R}^3$, we define the space $H_{tan}^s(\text{div}_S; S)$ by

$$H_{tan}^s(\text{div}_S; S) \ni u \longleftrightarrow u \in (H^s(S))^3, \ v \cdot u = 0 \text{ on } S, \ \text{div}_S u \in H^s(S),$$

---

[1] The Lipschitz boundary is enough for this fact.

where v is the outer unit normal to S, s is a real number, and div_S is the divergence on S associated with the induced Riemannian metric on S. Take a bounded domain $\Omega_{int} \subset \mathbf{R}^3$, and let $u = (u_E, u_H)$ be the solution of the following problem

$$\begin{cases} (\mathscr{H} - \lambda)u = 0 & \text{in} \quad \Omega_{int}, \\ v \times u_E = f_E \in H_{tan}^{-1/2}(\mathrm{div}_S; S), \end{cases} \tag{5.5}$$

where $S = \partial\Omega_{int}$. Let $\mathscr{H}_{int}$ be $\mathscr{H}$ with boundary condition $v \times u_E = 0$, and $\sigma_p(\mathscr{H}_{int})$ be its point spectrum. Then, for $\lambda \notin \sigma_p(\mathscr{H}_{int})$, (5.5) has a unique solution $u = (u_E, u_H)$. We define the interior E-M map by

$$\Lambda_{int}(\lambda) : H_{tan}^{-1/2}(\mathrm{div}_S; S) \ni f_E \to v \times u_H \in H_{tan}^{-1/2}(\mathrm{div}_S; S).$$

Assume that $\epsilon(x) - \epsilon_0 = 0$, $\mu(x) - \mu_0 = 0$ for $|x| > R - 1$, and take $S = \{|x| = R\}$.

Lemma 5.1 *The S-matrix $S(\lambda)$ determines the interior E-M map $\Lambda_{int}(\lambda)$.*

As in the case of Schrödinger operators, one can construct Faddeev's Green operator for the Maxwell equation in $\mathbf{R}^3$, and the Faddeev scattering amplitude. Ola, Päivärinta, and Sommersalo [154] solved the inverse boundary value problem by using Faddeev's Green operator. Combining these results, one can solve the inverse scattering problem for the Maxwell equation in the following way. First from the S-matrix, one constructs the interior E-M map. By using Faddeev's Green operator, one constructs a non-physical eigenfunction $u = (E, H)$ of the Maxwell operator, whose value on S is determined by the S-matrix. We then consider the integral on S:

$$(v \times E, M_0^*)_{L^2(S)} + (\Lambda_{int}(\lambda)v \times E, E_0^*)$$

where (E_0^*, M_0^*) is a suitable solution of the free Maxwell equation containing a large parameter. Letting it tend to infinity, one obtains the following non-linear elliptic equation:

$$\begin{cases} \Delta a + F(a, b) = pa, \\ \Delta b + F(b, a) = qb, \end{cases}$$

where $a = (\mu/\mu_0)^{1/2}$, $b = (\epsilon/\epsilon_0)^{1/2}$, $F(z_1, z_2) = \lambda(z_1 - z_1^2 z_2^2)$ and p, q are computed from the data. Ola, Päivärinta, and Sommersalo proved that this equation has a unique solution. We then have the following theorem.

Theorem 5.1 *Assume that $\epsilon(x)$, $\mu(x)$ are scalar functions. If $\epsilon(x) - \epsilon_0 \in C_0^\infty(\mathbf{R}^3)$ and $\mu(x) - \mu_0 \in C_0^\infty(\mathbf{R}^3)$, $\epsilon(x)$ and $\mu(x)$ are uniquely determined from the S-matrix $S(\lambda)$ of one fixed energy.*

The backscattering for the Maxwell equation was studied by Wang [192].

For the anisotropic case, i.e. when $\epsilon(x)$ and $\mu(x)$ are matrices, the inverse problem becomes much harder. The polarization independent case, i.e. $\epsilon(x) = k(x)\mu(x)$ for some scalar function $k(x) > 0$, the inverse boundary value problem was studied by Kurylev–Lassas–Sommersalo [125] and Kenig–Salo–Uhlmann [110]. Joshi and MacDowall [99] studied the chiral medium, i.e. $\mathscr{M} = \begin{pmatrix} a & c \\ -c & b \end{pmatrix}$.

The Maxwell equation is a good physical example of the equation for differential forms. As for the inverse spectral problem for differential forms, it is very hard to recover all information of the manifold from the spectral data. Belishev and Sharafutdinov [17] studied the reconstruction of the Betti number of manifolds from the D-N map. Krupchyk, Kurylev, and Lassas [123] proved that the dynamical response operator for the Maxwell equation given on a part of the boundary determines the Betti numbers. These works suggest that for the inverse spectral problem for differential forms, Betti number is a natural object.

Here, let us briefly recall the definition of Betti number adapted to the Maxwell equation. Let (M, g) be a three-dimensional compact Riemannian manifold with boundary. The space of harmonic fields is defined by

$$\mathscr{H}^k(M) = \{\omega \in L^2(M, \Lambda^k T^*(M)) \; ; \; d\omega = 0, \; d(*_g\omega) = 0\},$$

where Λ is the exterior product and $*_g$ is the Hodge star operator with respect to the metric g. Let $i^* : C^\infty(M, \Lambda^k T^*(M)) \to C^\infty(\partial M, \Lambda^k T^*(M))$ be the pull-back of the imbedding $i : \partial M \to M$. The space of Neumann harmonic fields is defined by

$$\mathscr{H}^k_N(M) = \{\omega \in \mathscr{H}^k(M) ; \; i^*(*_g\omega) = 0\},$$

which is isomorphic to the k th homology group of M. The k th *absolute Betti number* of M is then defined by

$$\beta_k(M) = \dim \mathscr{H}^k_N(M), \quad k = 0, 1, 2, 3.$$

For example, let $\Omega \subset \mathbf{R}^3$ be a bounded domain diffeomorphic to a ball glued with h handles. Then, $\beta_1(\Omega) = h$.

Now, let us consider the inverse scattering problem in an exterior domain $\Omega \subset \mathbf{R}^3$. Assume that $\partial\Omega = S_{int}$ is a compact smooth surface in $\mathbf{R}^3$ and there exists $R > 0$ such that for $|x| > R - 1$, $\epsilon(x) = \epsilon_0 I_3$, $\mu(x) = \mu_0 I_3$ for some constants $\epsilon_0, \mu_0 > 0$. On the exterior domain Ω, we consider the Maxwell equation and its S-matrix $S(\lambda)$.

Take a sphere $S_R = \{|x| = R\}$, and let Ω_{int} be the bounded domain with boundary $S_R \cup S_{int}$. Define the interior E-M map by $\nu \times E \to \nu \times H$ on S_R, where $u = (E, H)$ is the solution to the Maxwell equation $(\mathscr{H} - \lambda)u = 0$ in Ω_{int} with boundary condition

$$\nu \times E = f_E \quad \text{on} \quad S_R, \quad \nu \times E = 0 \quad \text{on} \quad S_{int}.$$

Then, one can show that $S(\lambda)$ determines the interior E-M map, which in turn determines the response operator R defined by $f \rightarrow H|_{S_R}$, where $u = (E, H)$ is the solution to the time-dependent Maxwell equation with boundary data $E = f$ on S_R and $E = 0$ on S_{int}. Krupchyk, Kurylev, and Lassas proved that this response operator determines the Betti number of Ω_{int}. Since the S-matrix $S(\lambda)$ for all $\lambda \in \mathbf{R} \setminus \{0\}$ determines the response operator, we have the following theorem.

Theorem 5.2 *The knowledge of the S-matrix $S(\lambda)$ for all $\lambda \in \mathbf{R} \setminus \{0\}$ determines the Betti numbers of the interior domain Ω_{int}.*

For example, the number of handles of the boundary $\partial \Omega$ is determined by the S-matrix of the Maxwell equation in the exterior domain Ω.

The above result is extended to non-compact Riemannian manifolds in the following way. Let $\mathcal{M}$ be a non-compact Riemannian manifolds of the form

$$\mathcal{M} = \mathcal{K} \cup \mathcal{M}_1 \cup \cdots \cup \mathcal{M}_N,$$

where $\mathcal{K}$ is a bounded open subset and each $\mathcal{M}_i$ is isometric to $\mathbf{R}^3 \cap \{|x| > R\}$. Let $\epsilon(x)$, $\mu(x)$ be positive definite symmetric tensors which are equal to identity on $\mathcal{M}_i$. The S-matrix for the Maxwell equation on $\mathcal{M}$ is then an $N \times N$ matrix operator. Fix $p \in \mathcal{K}$ arbitrarily, and consider a compact manifold with boundary $\mathcal{M}_{int} = \mathcal{M} \cap \{dist(x, p) < R\}$. Then, the knowledge of $S_{11}(\lambda)$, the $(1,1)$ entry of $S(\lambda)$, determines the Betti numbers of $\mathcal{M}_{int}$.

We need to mention here interesting results for the determination of the domain with uniform media. For the Helmholtz equation in an exterior domain, it is known that the whole domain is determined by the far-field pattern of all frequency. This is attributed to Schiffer [128], and the complete proof was given by Colton–Kress [44]. This result was further extended to a fixed frequency and for the Maxwell equation [45]. For polyhedral domains, only a finite number of incident waves with a fixed frequency is enough to determine the whole domain [133, 134].

For the details of contents in this section, see [88].

5.3   Perturbed Periodic Lattices

Many topics discussed in this book have their counterparts in discrete systems or graphs. For a survey of inverse problems in graph theory, see [22], especially its Chapter 7 and an abundance of references therein. The theory of isospectral graphs, inverse scattering on star-shaped graphs are developed in the same way as in the continuous model. The BC-method is also extended on graphs [18].

A remarkable fact is that inverse boundary value problems on planar graphs are solved including the characterization problem [41–43, 47–50]. In these works, we find a lot of geometrical similarities between continuous and discrete models. As we have seen in Sect. 2.4.2, the inverse boundary value problem on a compact Riemannian manifold is expected to be uniquely solved up to a diffeomorphism leaving the boundary invariant. In the case of planar graph, the role of this diffeomorphism is played by *elementary transformations*, which are operations to simplify the graph structure.

Lattice is a simple model to describe wave motions on periodic structures. There are two models. One deals with difference equations on vertices, called discrete graph, and the other studies differential equations on edges, called quantum graph. There is a rather general framework for the spectral theory and inverse scattering for Schrödinger operators on perturbed periodic lattices including square, triangular, hexagonal, diamond, Kagome lattices [9]. In particular, it deals with the graphene and graphite, which have important applications in nano-technology. As perturbations, for discrete graph one considers potentials on vertices or defects of vertices or edges, and for quantum graph potentials on edges and defects.

The interest of the spectral theory of discrete model or graph consists in the fact that we need both of one-dimensional and multi-dimensional results. Assuming that the perturbations are confined in a finite region, one can introduce the S-matrix, and prove that it determines uniquely the D-N map of the boundary value problem in a bounded domain.[2] By the results for the inverse boundary value problems for the planar graph, one can conclude that the S-matrix determines the perturbed graph up to elementary transformations [10].

In the discrete case, there exists a special type of solutions to the Helmholtz equation which vanishes in a half-space, and is not equal to 0 in the other half-space. It should be compared with the exponentially growing solution used in the continuous model. By using this solution, for the case of square, triangular, hexagonal lattices, one can reconstruct the local perturbation (potentials, defects of the lattice structure) from the knowledge of the S-matrix [10].

For the case of quantum graph with edge potentials which are symmetric with respect to the center of edge, using the above-mentioned solution, one can compute the Dirichlet eigenvalues of the potential. Then, by Theorem 1.12 (Borg's theorem), which is a starting point of the one-dimensional spectral theory, one can compute the edge potentials from the S-matrix [11].

5.4 Stäckel Metric

Let Ω be a domain in $\mathbf{R}^n$. A matrix-valued function $S(x) = (S_{ij}(x)) \in C^\infty(\Omega, GL_n(\mathbf{R}))$ is said to be a Stäckel matrix if $S_{ij}(x)$ depends only on x_i for all $1 \leq i \leq n$. A diagonal metric on Ω, $ds^2 = \sum_{i=1}^n g_{ii}(x)(dx_i)^2$, is said to be Stäckel metric if there exists a Stäckel matrix $S(x)$ and $1 \leq k \leq n$ such that $g^{ii}(x) = g_{ii}(x)^{-1} = S^{ki}(x)$ for all $1 \leq i \leq n$, where $(S^{ij}(x)) =$

---

[2]The Rellich type uniqueness theorem for the Schrödinger equation plays an important role here, to prove which HilbertNullstellensatz is used. We see here again an interesting interplay of spectral theory and elementary theory of function of several complex variables.

$S(x)^{-1}$. By a classical Theorem due to Stäckel [178], the Hamilton–Jacobi equation $\sum_{i=1}^{n} g^{ii}(x)(\partial W/\partial x_i)^2 = E$ is completely integrable by the separation of variables[3] if and only if ds^2 is a Stäckel metric. It is also known that under an additional assumption due to Robertson [171], the Stäckel metric is equivalent to the complete integrability of Helmholtz equation $-\Delta_g \phi = E\phi$ by the separation of variables.[4] This makes it possible to apply the method of separation of variables to inverse boundary value problems, which do not, apparently, admit the method of separation of variables. Using this fact and higher dimensional extension of Regge's method of complex angular momentum [168], Daudé, Karman, and Nicoleau [54] proved that the inverse boundary value problem can be solved for manifolds equipped with Stäckel metric. The application of this method to inverse scattering at a fixed energy is done by Gobin [66]. This class of metric is different from the ones discussed in Sect. 2.4.

---

[3]It means that $W(x)$ has the form $W(x) = \sum_{i=1}^{n} W_i(x_i)$.
[4]It means that ϕ is written as the product form $\phi = \prod_{i=1}^{n} \phi_i(x_i)$.

References

1. S. Agmon, Spectral properties of Schrödinger operators and scattering theory. Ann. Scoula Norm. Sup. Pisa **2**, 151–218 (1975)
2. S. Agmon, L. Hörmander, Asymptotic properties of solutions of differential equations with simple characteristics. J. d'Anal. Math. **30**, 1–30 (1976)
3. L.V. Ahlfors, *Complex Analysis* (McGrow-Hill, New York, 1966)
4. N.I. Akhiezer, I.M. Glazman, *Theory of Linear Operators in Hilbert Space, I, II* (Frederick Ungar Publishing Co., New York, 1961, 1963)
5. G. Alessandrini, J. Sylvester, Stability for a multidimensional inverse spectral theorem. Commun. Partial Differ. Equ. **15**, 711–736 (1990)
6. S. Alinhac, Non-unicité du problème de Cauchy. Ann. Math. **117**, 77–108 (1983)
7. V.A. Ambarzumian, Über eine Frage der Eigenwerttheorie. Z. Phys. **53**, 690–695 (1929)
8. W.O. Amrein, V. Georgescu, On the characterization of bound states and scattering states in quantum mechanics. Helv. Phys. Acta **46**, 635–658 (1973/74)
9. K. Ando, H. Isozaki, H. Morioka, Spectral properties of Schrödinger operators on perturbed periodic lattices. Ann. Henri Poincaré **14**, 347–383 (2013)
10. K. Ando, H. Isozaki, H. Morioka, Inverse scattering theory for Schrödinger operators on perturbed lattices. Ann. Henri Poincaré **19**, 3397–3455 (2018)
11. K. Ando, H. Isozaki, E. Korotyaev, H. Morioka, Inverse scattering on the quantum graph - Edge model for graphen (2019). arXiv:1911.05233[math-phy]
12. K. Astala, L. Päivärinta, Calderón's inverse conductivity problem in the plane. Ann. Math. **163**, 265–299 (2006)
13. K. Astala, L. Päivärinta, M. Lassas, Calderón's inverse problem for anisotropic conductivity in the plane. Commun. Partial Differ. Equ. **30**, 207–224 (2005)
14. R. Beals, R.R. Coifman, Multi-dimensional inverse scatterings and non-linear differential equations. Proc. Symp. Pure Math. **43**, 45–70 (1985)
15. M. Belishev, An approach to multidimensional inverse problems for the wave equation. Dokl. Akad. Nauk SSSR **297**, 524–527 (1987). Engl. transl. Soviet Math. Dokl. **36**, 481–484 (1988)
16. M. Belishev, V. Kurylev, To the reconstruction of a Riemannian manifold via its spectral data (BC-method). Commun. Partial Differ. Equ. **17**, 767-804 (1992)
17. M. Belishev, V. Sharafutdinov, Dirichlet to Neumann operator on differential forms. Bull. Sci. Math. **132**, 128–145 (2008)
18. M. Belishev, A. Vakulenko, Inverse problems on graphs: recovering the trees of strings by the BC-method. J. Inverse Ill-posed Probl. **14**, 29–46 (2006)
19. M. Bellassoued, M. Yamamoto, *Carleman Estimates and Applications to Inverse Problems for Hyperbolic Systems* (Springer, Berlin, 2017)

20. I. Beltita, A. Melin, Local smoothing for the backscattering transform. Commun. Partial Differ. Equ. **34**, 233–256 (2009)
21. P.H. Bérard, Variétés riemaniennes isospectrales nonisométriques. Séminaires Bourbaki **705**, 127–154 (1989)
22. G. Berkolaiko, P. Kuchment, *Introduction to Quantum Graphs*. Mathematical Surveys and Monographs, vol. 186 (American Mathematical Society, Providence, 2018)
23. A. Blagoveščenskii, A one-dimensional inverse boundary value problem for a second order hyperbolic equation. Zap. Nauchn. Sem. LOMI **15**, 85–90 (1969) (in Russian)
24. A. Blagoveščenskii, Inverse boundary problem for the wave propagation in an anisotropic medium. Trudy Mat. Inst. Steklova **65**, 39–56 (1971) (in Russian)
25. R.P. Boas, *Entire Functions* (Academic Press, Cambridge, 1954)
26. G. Borg, Eine Umkehrung der Sturm-Liouvilleschen Eigenwertaufgabe. Bestimmung der Differentialgleichung durch die Eigenwerte. Acta Math. **78**, 1–96 (1946)
27. R. Brooks, P. Perry, P. Petersen, Compactness and finiteness theorems for isospectral manifolds. J. Reine Angew. Math. **426**, 67–89 (1992)
28. J. Brüning, On the compactness of isospectral potentials. Commun. Partial Differ. Equ. **9**, 687–698 (1984)
29. J. Brüning, E. Heintze, Spektrale Starheit geweisser Drehlfächen. Math. Ann. **269**, 95–101 (1984)
30. A.L. Bukhgeim, Recovering the potential from Cauchy data in two dimensions. J. Inverse Ill-Posed Probl. **16**, 19–34 (2008)
31. A.L. Bukhgeim, M.V. Klibanov, Uniqueness in the large of a class of multidimensional inverse problems. Soviet Math. Dokl. **24**, 244–247 (1981)
32. A.P. Calderón, Uniqueness in the Cauchy problem for partial differential equations. Amer. J. Math. **80**, 1–36 (1958)
33. A.P. Calderón, On an inverse boundary value problem, in *Seminar on Numerical Analysis and Its Applications to Continuum Physics (Rio de Janeiro, 1980)* (Sociedade Brasileira de Matemática, Rio de Janeiro, 1980), pp. 65–73
34. T. Carleman, Sur un problème d'unicité pour les systèmes d'équations aux dérivées partielles à deux variables indépendants. Ark. Mat. Astron. Fys. **26**, 1–9 (1939)
35. K. Chadan, D. Colton, L. Päivärinta, W. Rundell, *An Introduction to Inverse Scattering and Inverse Spectral Problems* (SIAM, Philadelphia, 1997)
36. K. Chadan, P.C. Sabatier, *Inverse Problems in Quantum Scattering Theory* (Springer, Berlin, 1989)
37. S. Chanillo, A problem in electrical prospection and n-dimensional Borg-Levinson theorem. Proc. Amer. Math. Soc. **108**, 761–767 (1990)
38. S.J. Chapman, Drums that sound the same. Amer. Math. Monthly. **102**, 124–138 (1995)
39. S.Y. Cheng, S.T. Yau, On the regularity of the solution of the n-dimensional Minkowski problem. Comm. Pure Appl. Math. **29**, 495–516 (1976)
40. M. Choulli, P. Stefanov, Stability for the multi-dimensional Borg-Levinson theorem with partial spectral data. Commun. Partial Differ. Equ. **38**, 455–476 (2013)
41. Y. Colin de Verdière, Réseaux électriques planaires I. Comment. Math. Helv. **69**, 351–374 (1994)
42. Y. Colin de Verdière, *Cours Spécialisés 4, Spectre de Graphes* (Société mathématique de France, Paris, 1998)
43. Y. Colin de Verdière, I. de Gitler, D. Vertigan, Réseaux électriques planaires II. Commentarii Math. Helv. **71**, 144–167 (1996)
44. D. Colton, R. Kress, *Integral Equation Methods in Scattering Theory* (John Wiley and Sons, Hoboken, 1983)
45. D. Colton, R. Kress, *Inverse Acoustic and Electromagnetic Scattering Theory*, 3rd edn. (Springer, Berlin, 2013)
46. M.M. Crum, Associated Sturm-Liouville operators. Quart. J. Math. Oxford **6**, 121–127 (1955)
47. E.B. Curtis, J.A. Morrow, Determining the resistors in a network. SIAM J. Appl. Math. **50**, 918–930 (1990)

48. E.B. Curtis, J.A. Morrow, The Dirichlet to Neumann map for a resistor network. SIAM J. Appl. Math. **51**, 1011–1029 (1991)
49. E.B. Curtis, J.A. Morrow, *Inverse Problems for Electrical Networks* (World Scientific, Singapore, 2000)
50. E.B. Curtis, E. Mooers, J.A. Morrow, Finding the conductors in circular networks. Math. Modeling Numer. Anal. **28**, 781–813 (1994)
51. B.E. Dahlberg, E. Trubowitz, The inverse Sturm-Liouville problem III. Comm. Pure Appl. Math. **37**, 255–267 (1984)
52. G. Darboux, Sur une proposition relative aux équation linéaires. C. R. Acad. Sci. Paris **94**, 1456–1459 (1882)
53. T. Daudé, N. Karman, F. Nicoleau, Local inverse scattering at fixed energy in spherically symmetric asymptotically hyperbolic manifolds. Inverse Probl. Imaging **10**, 659–688 (2016)
54. T. Daudé, N. Karman, F. Nicoleau, The anisotropic Calderón problem on 3-dimensional conformally Stäckel manifolds (2019). J. Spectr. Theory. arXiv:1909.01669
55. D. Dos Santos Ferreira, C. Kenig, M. Salo, G. Uhlmann, Limiting Carleman weights and anisotropic inverse problems. Invent. Math. **178**, 119–171 (2009)
56. G. Eskin, J. Ralston, The inverse backscattering problem in three dimensions. Commun. Math. Phys. **124**, 169–215 (1989)
57. G. Eskin, J. Ralston, Inverse scattering problem for the Schrödinger equation with magnetic potential at a fixed energy. Commun. Math. Phys. **173**, 199–224 (1995)
58. G. Eskin, J. Ralston, E. Trubowitz, On isospectral periodic potentials in $\mathbf{R}^n$. Communications on Pure and applied Mathematics **35**, 647–676 (1984)
59. L.D. Faddeev, Increasing solutions of Schrödinger equation. Sov. Phys. Dokl. **10**, 1033–1035 (1966)
60. L.D. Faddeev, Factorization of the S-matrix for the multi-dimensional Schrödinger operator. Sov. Phys. Dokl. **11**, 209–211 (1966)
61. L.D. Faddeev, Inverse problem of quantum scattering theory. J. Sov. Math. **5**, 334–396 (1976)
62. G. Freiling, V. Yurko, *Inverse Sturm-Liouville Problems and their Applications* (Nova Science Publishers, Hauppauge, 2001)
63. I.M. Gel'fand, Some aspects of functional analysis and algebra, in *1957 Proceedings of the international Congress of Mathematics*, vol. 1 (North-Holland Publishing Co., Amsterdam, 1954), pp. 253–276
64. I.M. Gel'fand, B.M. Levitan, On the determination of a differential equation from its spectral function. Izvestiya Akad. Nauk SSSR. Ser. Mat. **15**, 309–360 (1951). English translation: Amer. Math. Soc. Transl. (2) **1**, 253–304 (1955)
65. F. Gesztesy, B. Simon, On local Borg-Marchenk uniqueness theorem. Comm. Math. Phys. **211**, 273–287 (2000)
66. D. Gobin, Inverse scattering at a fixed energy on three-dimensional asymptotically hyperbolic Stäckel manifolds. Publ. RIMS, Kyoto Univ. **54**, 245–316 (2018)
67. R. Goldberg, *Fourier Transforms* (Cambridge University Press, Cambridge, 1961)
68. A. Greenleaf, G. Uhlmann, Recovering of singularities of a potential from singularities of scattering data. Commun. Math. Phys. **157**, 549–572 (1993)
69. A. Greenleaf, M. Lassas, G. Uhlmann, On non-uniqueness for Caledron's inverse problem. Math. Res. Lett. **10**, 685–693 (2003)
70. A. Grigis and J. Sjöstrand, *Microlocal Analysis for Differential Operators, An Introduction*. London Mathematical Society Lecture Note Series, vol. 196 (Cambridge University Press, Cambridge, 1994)
71. C. Guillarmou, A. Sa Barreto, Scattering and inverse scattering on ACH manifolds. J. Reine Angew. Math. **622**, 1–55 (2008)
72. P. Hähner, *A periodic Faddeev type solution operator*, J. Diff. Equations, **128**, 300–308 (1996)
73. W. Heisenberg, Die "beobachtbaren Grössen" in der Theorie der Elemetarteilchen. Z. Phys. **120**, 513–538 (1943)
74. E. Hellinger, Neue Begründung der Theorie quadratischer Formen von unendlich vielen Veränderlichen. J. Reine Angew. Math. **136**, 210–271 (1909)

75. E. Holmgrem, Über Systeme von linearen partiellen Differentialgleichungen, Öfversigt af kongl. Vetenskaps-Akademiens Förhandlingr **58**, 91–103 (1901)

76. L. Hörmander, *The Analysis of Linear Partial Differential Operators II* (Springer, Berlin, 1983)

77. L. Hörmander, A uniqueness theorem for second order hyperbolic differential equations. Commun. Partial Differ. Equ. **17**, 699–714 (1992)

78. T. Ikebe, Eigenfunction expansions associated with the Scrödinger operators and their applications to scattering theory. Arch. Rat. Mech. Anal. **5**, 1–34 (1960)

79. A. Ikeda, On lens space which are isospectral but not isometric. Ann. scient. Éc. Norm. Sup. **13**, 3-3-315 (1980)

80. A. Ikeda, Y. Yamamoto, On the spectra of 3-dimensional lens specs, Osaka. J. Math. **16**, 447–469 (1979)

81. O. Imanuvilov, G. Uhlmann, M. Yamamoto, The Calderón problem with partial data in two dimensions. J. Amer. Math. Soc. **23**, 655–691 (2010)

82. E. Isaacson, E. Trubowitz, The inverse Strum-Liouville problem I. Comm. Pure Appl. Math. **36**, 767–783 (1983)

83. E. Isaacson, H. McKean, E. Trubowitz, The inverse Strum-Liouville problem II. Comm. Pure Appl. Math. **37**, 1–11 (1984)

84. V. Isakov, *Inverse Problems for Partial Differential Equations (1998)* (Springer, Berlin, 1998)

85. H. Isozaki, Some remarks on the multi-dimensional Borg-Levinson theorem. J. Math. Kyoto Univ. **31**, 743–753 (1991)

86. H. Isozaki, Multi-dimensional inverse scattering theory for Schrödinger operators. Rev. Math. Phys. **8**, 591–622 (1996)

87. H. Isozaki, Inverse scattering theory for Dirac operators. Ann. l'I. H. P. Physique Théorique **66**, 237–270 (1997)

88. H. Isozaki, *Maxwell Equation - Inverse Scattering in Electromagnetism* (World Scientific, Singapore, 2018)

89. H. Isozaki, E.L. Korotyaev, Global transformations preserving Sturm-Liouville spectral data. Russ. J. Math. Phys. **24**, 51–68 (2017)

90. H. Isozaki, E.L. Korotyaev, Inverse spectral theory and the Minkowski problem for the surface of revolution. Dyn. Partial Differ. Equ. **14**, 321–341 (2017)

91. H. Isozaki, Y. Kuryev, *Introduction to spectral theory and inverse problems on asymptotically Hyperbolic Manifolds*. Mathematical Society of Japan Memoires, vol. 32 (2014)

92. H. Isozaki, M. Lassas, Inverse scattering on non-compact manifolds with general metric (2020). arXiv:2004.06431

93. H. Isozaki, Y. Kurylev, M. Lassas, Forward and inverse scattering on manifolds with asymptotically cylindrical ends. J. Funct. Anal. **258**, 2060–2118 (2010)

94. H. Isozaki, Y. Kurylev, M. Lassas, Spectral theory and inverse problem on asymptotically hyperbolic orbifolds (2013). arXIv:1312.0421v1[math.AP]

95. H. Isozaki, Y. Kurylev, M. Lassas, Recent progress of inverse scattering theory on non-compact manifolds. Contemp. Math. **615**, 143–163 (2014)

96. H. Isozaki, Y. Kurylev, M. Lassas, Conic singularities, generalized scattering matrix, and inverse scattering on asymptotically hyperbolic manifolds. J. für Reine Angew. Math. **724**, 53–103 (2017)

97. C.G.J. Jacobi, Zur Theorier der Variationsrechnung und der Differentialgleichungen. J. Reine Angew. Math. **17**, 68–82 (1837)

98. F. John, On linear differential equations with analytic coefficients. Unique continuation of data. Comm. Pure Appl. Math. **2**, 209–253 (1949)

99. M.S. Joshi, S.R. MacDowall, Total determination of material parameters from electromagnetic boundary information. Pacific. J. Math. **193**, 107–129 (2000)

100. M. Kac, Can one hear the shape of a drum? Amer. Math. Monthly. **73**, 1–23 (1966)

101. P. Kargaev, E. Korotyaev, Inverse problem for the Hill operator, a direct approach. Invent. Math. **129**, 567–593 (1997)

102. P. Kargaev, E. Korotyaev, Erratum the inverse problem for the Hill operator, a direct approach. Invent. Math. **138**, 227 (1999)

103. A. Katchalov, Y. Kurylev, Multidimensional inverse problem with incomplete boundary spectral data. Commun. Partial Differ. Equ. **23**, 55–95 (1998)

104. A. Katchalov, Y. Kurylev, M. Lassas, *Inverse Boundary Spectral Problems*. Monographs and Surveys in Pure and Applied Mathematics, vol. 123 (Chapman and Hall/CRC, London, 2001)

105. T. Kato, S.T. Kuroda, The abstract theory of scattering. Rockey Mt. J. Math. **1**, 127–171 (1971)

106. T. Kato, *Perturbation Theory for Linear Operators* (Springer, Berlin, 1966)

107. O. Kavian, Y. Kian, E. Soccorsi, Uniqueness and stability results for an inverse spectral problem in a periodic wave guide. J. Math. Pures Appl. **104**, 1160–1189 (2015)

108. I. Kay, H.E. Moses, The determination of the scattering potential from the spectral measure function, I, II, III. Nuovo Cimento **2**, 917–961 (1955)

109. C.E. Kenig, J. Sjöstrand, G. Uhlmann, The Calderon problem with partial data. Ann. Math. **165**, 567–591 (2007)

110. C. Kenig, M. Salo, G. Uhlmann, Inverse problems for the anisotropic Maxwell equations. Duke Math. J. **157**, 369–419 (2011)

111. G.M. Khenkin, R.G. Novikov, The $\bar{\partial}$-equation in the multi-dimensional inverse scattering problem. Russ. Math. Surv. **42**, 109–180 (987)

112. Y. Kian, A multidimensional Borg-Levinson theorem for magnetic Schrödinger operators with partial spectral data. J. Spectr. Theory **8**, 235–269 (2018)

113. M.V. Klibanov, Inverse problems and Carleman estimates. Inverse Prob. **8**, 575–596 (1992)

114. K. Kodaira, The eigenvalue problem for ordinary differential equations of the second order and Heisenberg's theory of S-matrices. Amer. J. Math. **71**, 921–945 (1949)

115. R. Kohn, M. Vogelius, Determining conductivity by boundary measurements. Comm. Pure Appl. Math. **37**, 289–298 (1984)

116. P. Koosis, *Introduction to H_p Spaces* (Cambridge University Press, Cambridge, 1998)

117. E.L. Korotyaev, D.S. Chelkak, The inverse Strum-Liouville problem with mixed boundary conditions. St. Petersburg Math. J. **21**, 761–778 (2010)

118. M.G. Kreĭn, Determination of the density of an inhomogeneous string from its spectrum. Doklady Akad. Nauk. SSSR **76**, 345–348 (1951) (in Russian)

119. M.G. Kreĭn, On inverse problems for an inhomogeneous string. Doklady Akad. Nauk. SSSR **82**, 669–672 (1951) (in Russian)

120. M.G. Kreĭn, On the transfer function of a one-dimensional boundary value problem of the second order. Dokl. Akad. Nauk SSSR **88**, 405–408 (1953) (in Russian)

121. M.G. Kreĭn, On a method of effective solution of an inverse boundary value problem. Dokl. Akad. Nauk SSSR **94**, 987–990 (1954) (in Russian)

122. K. Krupchyk, L. Päivärinta, A Borg-Levinson theorem for higher order elliptic operators. Int. Math. Res. Not. **6**, 1321–1351 (2012)

123. K. Krupchyk, Y. Kurylev, M. Lassas, Reconstruction of Betti numbers of manifolds for anisotropic Maxwell and Dirac systems. Commun. Anal. Geom. **18**, 963–985 (2010)

124. S.T. Kuroda, Scattering theory for differential operators, I, II. J. Math. Soc. Japan **25**, 75–104, 222–234 (1973)

125. Y. Kurylev, M. Lassas, E. Sommersalo, Maxwell's equation with a polarization independent wave velocity: direct and inverse problems. J. Math. Pures Appl. **86**, 237–270 (2006)

126. M. Lassas, G. Uhlmann, On determining a Riemannian manifold from the Dirichlet-to-Neumann map. Ann. Sci. Ecole Norm. Sup. **34**, 771–787 (2001)

127. R. Lavine, A. Nachman, On the inverse scattering transform for the n-dimensional Schrödinger operators, in *Topics in Soliton Theory and Exactly Solvable Nonlinear Equations*, ed. by M. Ablowitz, B. Fuchssteiner, M. Kruskal, Oberwolfach (World Scientific, Singapore, 1986)

128. P.D. Lax, R.S. Phllips, *Scattering Theory* (Academic Press, Cambridge, 1967)

129. N. Levinson, The inverse Sturm-Liouville problem. Mat. Tidsskr. B. **1949**, 25–30 (1949)

130. B.M. Levitan, *Inverse Sturm-Liouville Problems* (Nauka, Moscow, 1984) (in Russian); English translation, CNU Science Press, Utrecht (1987)

131. B.M. Levitan, M.G. Gasymov, Determination of a differential equation by two of ts spectra. Russ. Math. Surv. **19**(2), 1–63 (1964)

132. J. Liouville, Mémoire sur les developpement des fonctions ou parties des fonctions en série dont les diverse terms sont assujettis à une même équations différentielle du second ordre, contenant un paramètre variable. J. Math. Pures Appl. **1**, 253–265 (1836)

133. H. Liu, J. Zou, Uniqueness in an inverse acoustic obstacle scattering problem for both sound-hard and sound-soft polyhedral scatters. Inverse Porb. **22**, 515–524 (2006)

134. H. Liu, M. Yamamoto, J. Zou, Reflection principle for the Maxwell equation and its applications to inverse electromagnetic scattering theory. Inverse Prob. **23**, 2357–2366 (2007)

135. J.J. Loeffel, On an inverse problem in potential scattering theory. Ann. Inst. Henri Poincare Sect. A **8**, 339–447 (1968)

136. V.A. Marchenko, Some problems in the theory of second order differential operators. Doklady Akad. Nauk SSSR **72**, 457–460 (1950) (in Russian)

137. V.A. Marchenko, Some problems in the theory of linear differential operators. Trudy Moskov. Mat. Obshch. **1**, 327–420 (1952) (in Russian)

138. V.A. Marchenko, *Sturm-Liouville Operators and Applications* (Naukova Dumka, Kiev, 1977) (in Russian); English translation, Birkhäuser, Basel (1986)

139. V.A. Marchenko, I.M. Ostrovski, A characterization of the spectrum of the Hill operators. Math. USSR Sb. **26**, 493–554 (1975)

140. A. Martin, Construction of the scattering amplitude from differential cross section. Nuovo Cimento **59A**, 131–151 (1969)

141. H.P. Mckean, I.M. Singer, Curvature and the eigenvalues of the Laplacian. J. Diff. Geom. **1**, 43–69 (1967)

142. A. Melin, The Faddeev approach to inverse scattering, in *Partial Differential Equations and Mathematical Physics*, ed. by L. Hörmander, A. Melin. Progress in Non-linear Differential Equations and Their Applications (Birkhäuser, Basel, 1996), pp. 226–245

143. R.B. Melrose, *Geometric Scattering Theory* (Cambridge University Press, Cambridge, 1995)

144. R. Melrose, G. Uhlmann, Generalized backscattering and the Lax-Phillips transform. Serdica Math. J. **34**, 355–372 (2008)

145. H.E. Moses, Calculation of the scattering potential from the reflection coefficients. Phys. Rev. **102**, 559–567 (1956)

146. E. Mourre, Absence of singular continuous spectrum for certain self-adjoint operators. Commun. Math. Phys. **78**, 391–408 (1981)

147. J.L. Mueller, S. Siltanen, *Linear and Nonlinear Inverse Problems with Practical Applications* (Computational Science and Engineering, SIAM, Philadelphia, 2012)

148. A. Nachman, Reconstruction from boundary measurements. Ann. Math. **128**, 531–576 (1988)

149. A. Nachman, Global uniqueness for a two-dimensional inverse boundary value problem. Ann. Math. **143**, 71–96 (1996)

150. A. Nachman, M. Ablowitz, A multidimensional inverse scattering method. Stud. Appl. Math. **71**, 243–250 (1984)

151. A. Nachman, J. Sylvester, G. Uhlmann, An n-dimensional Borg-Levinson theorem. Commun. Math. Phys. **115**, 595–605 (1988)

152. R. Novikov, Multidimensional inverse spectral problem for the equation $-\Delta\psi + (v(x) - Eu(x))\psi = 0$. Funkt. Anal. Prilozh. **22**(4), 11–22 (1988) (in Russian); Funct. Anal. Appl. **22**, 263–278 (1988)

153. R.G. Newton, *Scattering Theory of Waves and Particles*, 2nd edn. (Springer, Berlin, 1982)

154. P. Ola, L. Päivärinta, E. Sommersalo, An inverse boundary value problem in electrodynamics. Duke Math. J. **70**, 617–653 (1993)

155. P. Ola, L. Päivärinta, V. Serov, Recovering of singularities from backscattering in two dimensions. Commun. Partial Differ. Equ. **26**, 697–715 (2001)

156. L. Päivärinta, L. Serov, An n-dimensional Borg-Levinson theorem for singular potentials. Adv. Appl. Math. **29**, 509–520 (2002)

157. P.A. Perry, *Inverse spectral problems on compact Riemannian manifolds*. Lecture Notes in Physics, vol. 345. Schrödinger Operators, ed. by H. Holden, A. Jensen (Springer, Berlin, 1989)

158. A.V. Pogorelov, *The Minkowski Multidimensional Problem*, (Russian) Hilbert's fourth problem, Izdat (Nauka, Moscow, 1974), 79 pp. English translation by V. Oliker. Introduction by Louis Nirenberg. Scripta Series in Mathematics (V. H. Winston and Sons, Washington; Halsted Press, Ultimo; John Wiley and Sons, New York, 1978)

159. P. Pöschel, E. Trubowitz, *Inverse Spectral Theory* (Academic Press, Boston, 1987)

160. A.Y. Povzner, On the expansion of an arbitrary function in terms of the eigenfunction of the operator $-\Delta u + cu$. Mat. Sbornik **32**, 109–156 (1953)

161. R.T. Prosser, Formal solutions of inverse scattering problems III. J. Math. Phys. **21**, 2648–2653 (1980)

162. Rakesh, M. Salo, Fixed angle scattering for almost symmetric or controlled perturbations (2019). arXiv:1905.03974v1[math.AP]

163. Rakesh, M. Salo, The fixed angle scattering problem and wave equation inverse problems with two measurements. Inverse Prob. **30**, 035005 (2020)

164. A. Ramm, Property C with constraints and inverse spectral problems with incomplete data. J. Math. Anal. Appl. **180**, 239–244 (1993)

165. A.G. Ramm, An inverse scattering problem with part of the fixed energy phase shifts. Comm. Math. Phys. **207**, 231–247 (1999)

166. A. Ramm, *Inverse Problems – Mathematical and Analytical Techniques and Applications to Engineering* (Springer, Berlin, 2004)

167. M. Reed, B. Simon, *Methods of Modern Mathematical Physics* (Academic Press, Cambridge, 1975)

168. T. Regge, Introduction to complex orbital momenta. Nuovo Comento **14**, 951–976 (1959)

169. J.M. Reyes, A. Ruiz, Reconstruction of the singularities of a potential from backscattering data in 2D and 3D. Inverse Prob. Imaging **6**, 321–355 (2012)

170. L. Robbiano, Théorème d'unicité adapté au contrôle des solutions des problémes hyperboliques. Commun. Partial Differ. Equ. **16**, 789–800 (1991)

171. H.P. Robertson, Bemerkung über separierbare Systeme in der Wellenmechanik. Math. Ann. **98**, 749–752 (1927)

172. D. Ruelle, A remark on bound states in potential-scattering theory. Nuovo Cimento A (10) **61**, 655–662 (1969)

173. A. Ruiz, Recovery of the singularities of a potential from fixed angle scattering data. Commun. Partial Differ. Equ. **26**, 1721–1738 (2001)

174. A. Sá Barreto, Radiation fields, scattering and inverse scattering on asymptotically hyperbolic manifolds. Duke Math. J. **129**, 407–480 (2005)

175. Y. Saito, *Spectral Representations for Schrödinger Operators with Long-Range Potentials*. Lecture Notes in Mathematics, vol. 727 (Springer, Berlin, 1979)

176. M. Salo, The Calderón problem on Riemannian manifolds, in *Inside Out II*, ed. by G. Uhlmann (MSRI Publications, Cambridge University Press, Cambridge, 2012), pp. 167–247

177. B. Simon, A new approach to inverse spectral theory, I. Fundamental formalism. Ann. Math. **150**, 1029–1057 (1999)

178. P. Stäckel, Über die Bewegung eines Punktes in einer n-fachen Mannigfaltigkeit. Math. Ann. **42**, 537–563 (1893)

179. P. Stefanov, Generic uniqueness for two inverse problems in potential scattering. Commun. Partial Differ. Equ. **17**, 55–68 (1992)

180. M.H. Stone, *Linear Transformations in Hilbert Space* (American Mathematical Society Colloquium Publications, New York, 1932)

181. J.C.F. Sturm, Mémoire sur les équations différentielles linéaires du second ordre. J. Math. Pures Appl. **1**, 106–186 (1836)

182. J.C.F. Sturm, Mémoire sur une classe d'équations à différences partielles. J. Math. Pures Appl. **1**, 373–444 (1836)

183. T. Sunada, Riemannian coverings and isospectral manifolds. Ann. Math. **121**, 169–186 (1985)

184. J. Sylvester, G. Uhlmann, A global uniqueness theorem for an inverse boundary value problem. Ann. Math. **125**, 153–169 (1987)
185. D. Tataru, Unique continuation for solutions to PDEs, between Hörmander's theorem and Holmgren's theorem. Commun. Partial Differ. Equ. **20**, 855–884 (1995)
186. A.N. Tikhonov, On uniqueness of the solution of an electroreconnaissance problem. Dokl. Akad. Nauk SSSR **69**, 797–800 (1949)
187. E.C. Titchmarsh, *Eigenfunction Expansions Associated with Second-order Differential Equations* (Oxford University Press, Oxford, 1946)
188. G. Uhlmann, A time-dependent approach to the inverse backscattering problem. Inverse Prob. **17**, 703–716 (2001)
189. G. Uhlmann, Inverse problems: seeing the unseen. Bull. Math. Soc. **4**, 209–279 (2014)
190. H. Urakawa, Bounded domains which are isospectral but not congruent. Ann. scient. Éc. Norm. Sup. **15**, 441–456 (1982)
191. H. Urakawa, Spectral of discrete and continuous Laplacians on graphs and Riemannian manifolds. Interdiscip. Inf. Sci. **3**, 95–105 (1997)
192. J.N. Wang, Inverse backscattering problem for Maxwell's equation. Math. Meth. App. Sci. **21**, 1441–1465 (1998)
193. C. Weber, A local compactness theorem for Maxwell's equations. Math. Meth. Appl. Sci. **2**, 12–15 (1980)
194. R. Weder, Generalized limiting absorption method and inverse scattering theory. Math. Meth. Appl. Sci. **14**, 509–524 (1971)
195. H. Weyl, Über gewöhnliche Differentialgleichungen mit Singularitäten und die zugehörigen Entwicklungen willkürlicher Funktionen. Math. Ann. **68**, 220–269 (1910)
196. J.A. Wheeler, On the mathematical description of light nuclei by the method of resonating group structure. Phys. Rev. **52**, 1107–1122 (1937)
197. D.R. Yafaev, *Mathematical Scattering Theory - Analytic Theory*. Mathematical Surveys and Monographs, vol. 158 (AMS, Providence, 2010)
198. K. Yosida, *Functional Analysis* (Springer, Berlin, 1966)
199. S. Zelditch, Kuznecov sum formulae and Szegö limit formulae on manifolds. Commun. Partial Differ. Equ. **17**, 221–260 (1992)
200. S. Zelditch, The inverse spectral problem for surfaces of revolution. J. Diff. Geom. **49**, 207–264 (1998)
201. M. Zworski, *Semiclassical Analysis*. Graduate Studies in Mathematics, vol. 138 (American Mathematical Society, Providence, 2012)

Index

© The Author(s), under exclusive licence to Springer Nature Singapore Pte Ltd. 2020
H. Isozaki, *Inverse Spectral and Scattering Theory*, SpringerBriefs
in Mathematical Physics 38, https://doi.org/10.1007/978-981-15-8199-1

Printed in the United States
By Bookmasters